Environment, Forced Migration and Social Vulnerability

Tamer Afifi · Jill Jäger
Editors

Environment, Forced Migration and Social Vulnerability

Springer

Editors
Dr. Tamer Afifi
United Nations University
Inst. Environment
Human Security (UNU-EHS)
Hermann-Ehlers-Str. 10
53113 Bonn
Germany
afifi@ehs.unu.edu

Dr. Jill Jäger
Independent Scholar
Arbeiterstrandstrasse 61
1210 Vienna
Austria
jill.jaeger@seri.at

ISBN 978-3-642-12415-0 e-ISBN 978-3-642-12416-7
DOI 10.1007/978-3-642-12416-7
Springer Heidelberg Dordrecht London New York

Library of Congress Control Number: 2010930718

Cover design: deblik, Berlin

Printed on acid-free paper

Springer is part of Springer Science+Business Media (www.springer.com)

Foreword

This book is one of the outputs of the conference on 'Environmental Change, Forced Migration, and Social Vulnerability' (EFMSV) held in Bonn in October 2008.

Migration is one of the oldest adaptation measures of humanity. Indeed, without migration the multitude of civilizations and interactions between them – peaceful and otherwise – would be hard to imagine. The United Nations (UN)-led global dialogue on migration is a clear sign that governments and the specialized UN agencies and bodies have recognized the need to view, govern, manage, and facilitate migration; to mitigate its negative effects; and to capitalize on the positive ones.

It is a common expectation among experts that environmentally induced migration will further increase in the decades to come. Hence, next to the political, economic, ethnic, social, financial, humanitarian, and security aspects of migration, the environmental component should urgently be considered in the ongoing international dialogue on migration. This need is also a challenge. Without appropriate scientific knowledge, assessment, definitions, and classifications, the intergovernmental frameworks would not be able to deal with these complex phenomena. The Five-Pronged-Approach as formulated by the United Nations University (UNU) may serve as a framework to identify the additional dimensions of this challenge next to – and actually simultaneously with – the scientific one. The five complementary measures – awareness raising, legal framework of recognition, empowering the humanitarian organizations to assist those on the move, institutional capacity development to drain the criminal component of the often illegal migration flows, and more scientific investigation – should be addressed together.

The EFMSV Conference is therefore part of a process in this multi-faceted approach. The goal of the conference was to reach a better understanding of the process of internal displacements, rural–rural and rural–urban migration, resettlements, and trans-boundary and transcontinental migration.

The study of environmentally induced migration, while having a 'history' of about three decades, experiences a recent renaissance. The International Organization for Migration (IOM)–United Nations Population Fund (UNFPA) International Dialogue on Migration in Bangkok, February 2007, could be mentioned in this regard as a benchmark event. UNU's publication *Control, Adapt or Flee: How to Face Environmental Migration?* (Spring 2007) summarizes the state of the art at that time. To further specify the research agenda, a process started in April 2008 with the UNU–IOM–United Nations Environment Programme (UNEP) Research Workshop on Migration and the Environment supported by the Munich Re Foundation (MRF) and the Rockefeller Foundation. This was continued by the enthusiastic work of 25 PhD candidates formulating the 'Hohenkammer Challenge' in July 2008 during the third MRF–UNU-EHS (United Nations University Institute for Environment and Human Security) Summer Academy on Social Vulnerability. The process towards the establishment of the Climate Change, Environment and Migration Alliance (CCEMA) by these and other partners continued at the EFMSV Conference in Bonn.

EFMSV was both an open conference and also an element of the European Commission funded Environmental Change and Forced Migration Scenarios (EACH-FOR) Project, implemented by 7 institutional members of the project consortium and receiving substantial assistance from IOM, the Organization for Security and Co-operation in Europe (OSCE), and many NGOs and universities in 23 case study countries. EACH-FOR was a unique project where common principles and methodologies were used to trace the 'environmental signal' in ongoing migration and potential decisions to move. The EFMSV Conference provided an opportunity to showcase the results of this ground-breaking project (see Part V).

I wish to thank all of the authors for their contributions to this book, as well as acknowledging the work of the Scientific Committee in reviewing the papers submitted to the conference and thus participating in the selection process for the book. Furthermore, I would like to thank the many organizations that helped to make the EMFSV Conference possible through their support, including the Robert Bosch Foundation, the Foundation for International Dialogue of the Savings Bank in Bonn, the German Research Foundation, the City of Bonn, the German foreign language broadcasting corporation (Deutsche Welle), the Deutsche Post, and the European Commission's Sixth Framework Programme for Research. We appreciate the fruitful collaboration with our EACH-FOR partners and partner organizations IOM, the Munich Re Foundation, OSCE, the UN Convention to Combat Desertification (UNCCD), UNEP, and a broad network of academic and policy professionals. The strong, growing community is a clear sign that migration, and within it the environmentally induced migration, is seen by many as one of the key challenges of the years to come.

Bonn, Germany Janos J.Bogardi

Preface

In recent decades much research has been done about environment and migration by separate communities of scholars. Over the past 5 years, however, there has been increasing collaboration between the experts, so that a more integrated view of the linkages between environmental change and migration is becoming available. This book, consisting of a set of papers presented at the Environmental Change, Forced Migration, and Social Vulnerability (EFMSV) Conference, held in Bonn in October 2008, illustrates some of the results of recent research.

The chapters in this book were selected on the basis of a peer-review process carried out by the Scientific Committee of the conference. A further selection of papers will be published in a special edition of the peer-reviewed journal *International Migration* focusing on migration and the environment.

We have divided the chapters of the book into five parts, although the reader will quickly note that most papers would be relevant for more than one of the topics. Overall, the chapters demonstrate the current state of the art in the study of the complex relationships between environmental change and migration, the implications of some of the issues of definitions, and the policy responses that are called for.

We would like to thank all of the authors for their prompt and uncomplicated responses to our requests for input. It has taken longer to get this book published than we originally anticipated, so we also thank them for their patience.

Bonn, Germany — Tamer Afifi
Vienna, Austria — Jill Jäger

Contents

Contributors

Jypara Abarikova Regional Center for Migration and Refugees, Bishkek, Kyrgyzstan, rcmr@mail.ru

Aigoul Abdoubaetova Anthropology Research Center Jorgo, Bishkek, Kyrgyzstan, a.abdoubaetova@osce-academy.net

Mehrigul Ablezova Social Research Center, American University of Central Asia, Bishkek, Kyrgyzstan, ablezovam@mail.auca.kg

Tamer Afifi Environmental Migration, Social Vulnerability and Adaptation Section, United Nations University Institute for Environment and Human Security (UNU-EHS), Bonn, Germany, afifi@ehs.unu.edu

Stefan Alscher Center on Migration, Citizenship and Development (COMCAD), University of Bielefeld, Bielefeld, Germany, stalscher@migra.org

Jörn Birkmann United Nations University Institute for Environment and Human Security, Section on Vulnerability Analysis, Bonn, Germany, birkmann@ehs.unu.edu

Frauke Bleibaum University of Bielefeld, Bielefeld, Germany, frauke.bleibaum@no-log.org

Hassan Bousetta Centre for Ethnic and Migration Studies (CEDEM), University of Liege, Liege, Belgium, Hassan.Bousetta@ulg.ac.be

Robin Bronen Resilience and Adaptation Program, University of Alaska, Fairbanks, AK, USA, rbronen@yahoo.com

Bruce Burson Refugee Status Appeals Authority, Auckland, New Zealand, bruceburson@hotmail.com

Nishara Fernando United Nations University Institute for Environment and Human Security, Section on Vulnerability Analysis, Bonn, Germany, nishara2000@yahoo.com

Richard Filčák Institute for Forecasting Studies, Slovak Academy of Sciences, Bratislava, Slovak Republic, rfilcak@yahoo.co.uk

Johannes Frühmann Sustainable Europe Research Institute, Vienna, Austria, johannes.fruehmann@seri.at

François Gemenne Centre for Ethnic and Migration Studies (CEDEM), University of Liege, Liege, Belgium; Centre for International Studies and Research (CERI), Sciences Po, Paris, France, francois.gemenne@sciences-po.org

Jill Jäger Sustainable Europe Research Institute, Vienna, Austria, jill.jaeger@seri.at

Zeynep Kadirbeyoglu Bogazici University, Istanbul, Turkey, kadirbey@yahoo.com

Marco Martiniello Centre for Ethnic and migration studies (CEDEM), University of Liege, Liege, Belgium, M.Martiniello@ulg.ac.be

Yojana Miner Fuentes Coordinadora Nacional para la Reducción de Desastres, CONRED, Guatemala, Guatemala, yojaminer@yahoo.com

Emil Nasritdinov Social Research Center, American University of Central Asia, Bishkek, Kyrgyzstan, emilzn@gmail.com

Wim Naude World Institute for Development Economics Research, United Nations University, Helsinki, Finland, Wim@wider.unu.edu

Etienne Piguet Institute of Geography, University of Neuchâtel, Neuchâtel, Switzerland, Etienne.piguet@unine.ch

Alice Poncelet Centre for Ethnic and Migration Studies (CEDEM), University of Liege, Liege, Belgium, poncelet.alice@gmail.com

Kaushalya Ramachandran Central Research Institute for Dryland Agriculture, Hyderabad, India, kausalyar@yahoo.com

Kerstin Schmidt-Verkerk Sussex Centre for Migration Research, University of Sussex, Brighton, UK, k.schmidt-verkerk@sussex.ac.uk

Padmaja Susarla Department of Geography, Science College, Osmania University, Hyderabad, India, padmaja45@rediffmail.com

Petra Tschakert Department of Geography, Pennsylvania State University, University Park, IL, USA, petra@psu.edu

Raymond Tutu Department of Geography, Pennsylvania State University, University Park, IL, USA, rat154@psu.edu

Juan Carlos Villagrán de León Risk Management Section, United Nations University Institute for Environment and Human Security, Bonn, Germany, juan-carlos.villagran@unoosa.org

Chloe Anne Vlassopoulos CURAPP-CNRS/University of Picardie Amiens, France, chloe.vlassopoulou@u-picardie.fr

Koko Warner United Nations University Institute for Environment and Human Security, Section on Environmental Migration, Social Vulnerability, and Adaptation, Bonn, Germany, warner@ehs.unu.edu

Part I
Defining the Terms: Introduction

Environment, Forced Migration, and Social Vulnerability are highly interrelated with each other, since when dealing with 'Environmental Migration', the mechanism through which environmental problems lead to forced or voluntary migration is usually social vulnerability. Whether we refer to 'Environmental Migrants', 'Environmentally Induced Migrants', or any terms covered in the definition tree that is discussed in more detail within the "Bonn Points", it is important to avoid using 'Environmental Refugee', since using this expression could damage the prospects of 'Political Asylum Seekers' who might be subject to death or torture if they are not granted refugee status.

The Geneva Refugee Convention of 1951 does not recognise people fleeing from environmental problems as refugees, since the Convention is constrained by specific criteria that create the refugee status, such as race, colour, religion, and ideology. Moreover, the lack of concrete data about the number of current and potential environmental migrants is an obstacle that hinders many institutions from taking serious measures that help mitigate environmental degradation, prevent environmental migration, decrease the number of environmental migrants, or assist this category of people in the countries/regions of destination.

This part of the book focuses on the recognition of environmentally displaced persons. The first chapter explores possible pathways for recognition under the Geneva Refugee Convention of people for whom environmental issues form part of the migratory push factor. This is followed by a chapter that discusses the institutional barriers to the recognition and assistance of environmentally forced migrants and suggests that a more pragmatic approach using available policy instruments would be the best next step in overcoming the barriers. Finally, the third chapter in this section discusses the migration in the case of Hurricane Katrina in New Orleans and highlights that the people who were forced to move objected strongly to being called 'refugees' in the media, since they had not left their home country.

Chapter 1
Environmentally Induced Displacement and the 1951 Refugee Convention: Pathways to Recognition

Bruce Burson

1 Introduction

The potential for environmental degradation to cause or contribute to migration outflows raises the issue of the appropriate response by receiving states and the international community in general. As one of the primary international treaties dealing with migration, there is a need for the 1951 United Nations Convention Relating to the Status of Refugees (the Convention) to be placed at the heart of any debate about the recognition of environmentally displaced persons. The question is, In what way? While some (Conisbee and Simms, 2003: 30) have called for the Convention to be amended to include the environmentally displaced, the chances of this happening are slim. The recent trend in refugee-receiving states in the global north[1] has been towards raising barriers to entry through interdiction and truncated legal process upon arrival. Moreover, refugee-receiving states in the global south, which already bear a disproportionate burden in hosting existing refugee populations, are unlikely to agree to assuming additional responsibilities. Another way may be to use the Convention as a basis to secure a parallel international instrument covering 'environmental refugees' in much the same way as a parallel agreement deals with stateless persons. This chapter does not address this issue, although I broadly

---

B. Burson (✉)
Refugee Status Appeals Authority, Auckland, New Zealand
e-mail: bruceburson@hotmail.com

This chapter is written in the author's personal capacity. It neither does purport to represent the views of the New Zealand Refugee Status Appeals Authority or any other member thereof, nor does it purport to represent the view the author would take in any particular case.

[1] Betts (2005: 4) observes

> In the refugee context, the 'north' can be regarded to comprise the industrialised third-country asylum states, which are generally outside the refugees' regions of origin. Meanwhile, the 'south' invariably comprises the refugee producing, transit or first asylum host states within regions of origin.

T. Afifi, J. Jäger (eds.), *Environment, Forced Migration and Social Vulnerability*,
DOI 10.1007/978-3-642-12416-7_1, © Springer-Verlag Berlin Heidelberg 2010

agree with Renaud et al. (2007: 29) that there is an urgent need for an appropriate policy response at the international level.

This chapter explores a third avenue – that of possible pathways into the Convention for persons for whom environmental issues form part of the migratory 'push' factor. The first section considers literature on environmentally displaced persons. It delves into literature on linkages between environmental degradation, migration, and conflict, to identify potential pathways into the Convention. The second section considers the Convention's Article 1A(2) definition in more detail – something generally absent from the literature. It identifies some jurisprudential issues that may arise in dealing with claims under the Convention by the environmentally displaced in order to signpost some issues that may arise for refugee decision-makers, advocates, and policymakers.

2 The 1951 Refugee Convention and Environmental Refugees

2.1 Conceptualising the Term

Distinctions are drawn in the literature between 'migrants' – who respond to both push and pull factors – and 'environmental refugees' – who have no choice but to flee (Lee, 1997: 275; Renaud et al., 2007: 29; Suhkre, 1994: 482). Yet the term 'environmental refugee' is contentious. Indeed, some argue that the term 'environmental refugee' should be avoided altogether (Castles, 2002: 5; Hugo, 1996: 109). When used, differing conceptualisations of an 'environmental refugee' exist (Hugo, 1996: 107; Lonergan and Swain, 1999: 2; Suhkre, 1994: 482). These wider definitions share some common ground in that they dispense with any requirement for international migration, but, beyond this, differ greatly (Renaud et al., 2007: 13). Typically, however, there is a distinction between wider conceptions – such as those by Olson (1979: 130),[2] El-Hinnawi (1985: 4) and Jacobsen (1988: 37–38) – and the 'narrow' definition of refugees in the Convention.

Suhkre (1994: 482) argues that different theoretical frameworks underpin these conceptions and contrasts the narrow 'legal' conception in the Convention with wider 'sociological' categorisations in which a refugee is understood as no more than a person who is forced to flee involuntarily. Suhkre (1994: 488) goes on to argue that, while persons who are compelled to involuntarily migrate simply to survive may meet the sociological definition of a refugee, they may not necessarily meet an essential element of the legal definition: the need for protection from the state. Similarly, Renaud et al. (2007: 14) argue that the 'strict' Convention definition 'does not appear to be suitable for describing those displaced by environmental factors'. They point to difficulties this category of migrant face in establishing that the environment can constitute an agent of persecution with an intent to harm.

[2] Cited in Hugo (1996: 107).

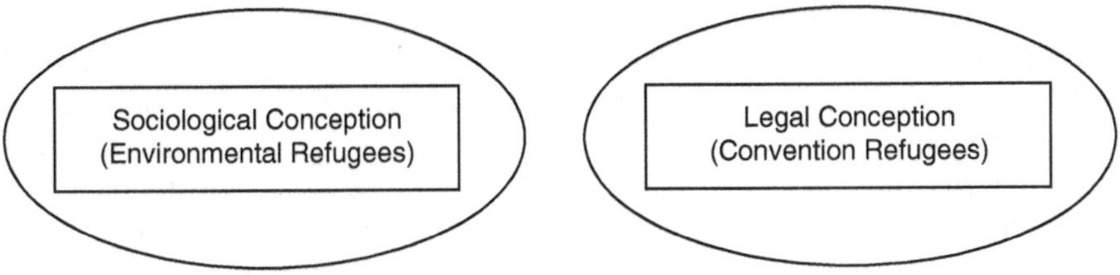

Fig. 1.1 The divide between legal and sociological conceptions of refugees

This notion of a divide between sociological and legal conceptions is shown in Fig. 1.1.

The difficulty with this view of the relationship is that it underemphasises the fact that environmentally induced migration does not occur in a vacuum. It takes place within a particular societal setting – a setting which may have features which can create a pathway into the Convention. To understand what those features are, we need to look at literature on environmental degradation, conflict, and migration.

2.2 Environmental Degradation, Conflict, and Migration

Reflecting the emergence of environmental issues as a discreet aspect of international relations and international law, since the 1970s, increasing attention has been given to the linkage between environmental issues and both armed conflict and security (Gleditsch, 1998: 381; Ronnfelt, 1997: 473). The work of the so-called Toronto Group (Homer-Dixon and Percival, 1996: 6–10) concluded, inter alia, that there was a positive linkage between environmental scarcity, conflict, and migration. In particular, if social and economic adaptation is unsuccessful, environmental scarcity contributes to impoverishment and migration. It may sharpen distinctions between groups and enhance the opportunity for participation in violent conflict.

This work has proved to be highly controversial.[3] It is not a function of this chapter to enter into this debate. For present purposes, two important points emerge from the literature. First, there appears to be some general acceptance that environmental issues *can* pose threats to security and induce violent conflict and migration, albeit in a highly uncertain manner and through complex social and political processes (Baechler, 1999: 107; Barnett, 2003: 7; Black, 2001: 3; Castles, 2002: 5; Gleditsch, 1998: 389; Levy, 1995: 59; Lonergan and Swain, 1999: 8; Richmond, 1993: 8; Ronnfelt, 1997: 478; Suhkre, 1997: 257; Volger, 2002: 180).

Second, if environmental issues are to play some role in generating conflict, this is likely to occur along a number of sub-state fault-lines, not at an

[3] See, for example, Gleditsch (1998: 390–391); Levy (1995: 57); Ronnfelt (1997: 477–478).

international level (Ashton, 2002: 240; Baechler, 1999: 109; Barnett, 2003: 10). Ashton has developed a useful five-level typology of potential conflict, four of which operate at a sub-state or domestic level, namely, intra-community; inter-community; inter-provincial; and urban–rural conflict. That any conflict arising from environmental degradation is likely to be at a sub-state level is unsurprising. In every year since the end of the Second World War, the number of ongoing internal armed conflicts has exceeded the number of interstate conflicts (Harbom and Wallensteen, 2005: 627).

This multiplicity of possible sub-state level conflict fault-lines is important. If a state becomes the site of conflict, safety may well be seen by potential victims to lie outside its borders. This is significant, as the Convention requires that the claimant be outside their country of nationality or former habitual residence. In broad terms, this multiplicity increases the potential for migration across borders to take place, particularly where an affected group has minority status. Studies conducted in the aftermath of famine and other natural disasters provide evidence of the political weighting of state response in which the recovery needs of marginalised groups are not met (Pelling and Dill, 2006: 5; Raleigh et al., 2008: 33). Should environmental degradation be of sufficient magnitude, such marginalisation may mean that external migration becomes a more attractive option, particularly where majority group communities are also affected by the same environmental event in a state with limited capacity to respond. A recent report (Baird, 2008: 10–11) observes that, in many cases, the impact of climate change is felt most acutely by minority and indigenous communities which often live in the most marginal areas. If these communities have historically encountered entrenched official and societal discrimination, this may result in a failure to respond to their needs in the wake of environmental disasters caused or exacerbated by climate change.

It is also important to note that these various types of sub-state conflicts need not constitute a civil war in order for the Convention to be potentially engaged; indeed, it may be more difficult to establish a successful refugee claim in a civil war context. Although other legal hurdles will have to be crossed, it is important to stress that localised environmental issues leading to a less intense level of conflict can still potentially ground a claim for Convention-based recognition.

2.3 The Relationship Reconceptualised

The existence of complex relationships between environmental degradation, conflict, and migration means that it is an oversimplification to insist on immutable distinctions between environmentally displaced persons and those eligible for recognition as Convention refugees. Given the potential for issues of environmental degradation to collide with issues of poverty, inequality, discriminatory modes of governance, and human rights violations, the environmental issue may acquire attributes which potentially bring affected persons within the scope of the Convention. If so, there is no principled basis for distinguishing between these claims and claims where such a link is absent in terms of recognition.

The reality is that environmental degradation does not occur in a vacuum, isolated from anthropogenic influence. Environmental degradation is intimately bound up with long-term issues of development, population growth, and economic and social policy choices (Hugo, 1996: 109; Renaud et al., 2007: 22; Vlachos, 1997: 297, 305). This is particularly true in relation to climate change (IPCC, 2007: 9). This historical context, when mixed with activity of a discriminatory nature, can in principle produce environmentally displaced persons who meet the Convention's definition. While in such cases the presence of these other discriminatory practices is essential for Convention-based recognition, this does not mean that the underlying environmental issue ceases to have all relevance. Historical policy choices, particularly development and infrastructure-related policy choices (Turton, 2003: 11), may evidence a pattern of official discrimination – a relevant consideration for the Convention inquiry. Furthermore, that the impacts of environmental degradation may not be transient in nature may assist in establishing that there is no real chance that things will improve – a key factor in establishing Convention recognition.

In other words, the legal conception of refugee as set out in the Convention is better conceptualised, not as alternative to, but rather as a subset of the sociological conception as represented in Fig. 1.2.

The arrows in Fig. 1.2 reflect the idea that the relative size of that group within the sociological conception who would also meet the legal conception as set out in the Convention is variable. In some instances, the gap between the legal and sociological conceptions might be very small. Where environmental degradation is used as a direct weapon of repression against an entire section of the population, such as the Iraqi Marsh Arabs post-1991 (Human Rights Watch, 2003: 4), the numbers of people who may fall within the Convention will be greater. At the other end of the spectrum may be cases where it is only certain individuals within a larger group of environmentally displaced persons whose circumstances may also come within the legal conception. Examples here could include community leaders or activists who seek redress or agitate against the state activity which caused or contributed (or at least is perceived to have contributed) to the environmental degradation in the first place or a discriminatory state response in the wake of some unplanned and unintended environmental disaster.[4] Events in Burma following Cyclone Nargis in May 2008 illustrate the point. According to Human Rights Watch (2009: 226) the authorities in Burma, a country with an extremely poor human rights record, arrested several prominent political prisoners for their role in cyclone relief activities including one who had criticised the ruling State Peace and Development Council's relief efforts in interviews with foreign media.

This conceptualisation more accurately reflects what has come to be termed 'complex displacement' – situations where environmental degradation and

[4] As to which, see Pelling and Dill (2006: 5).

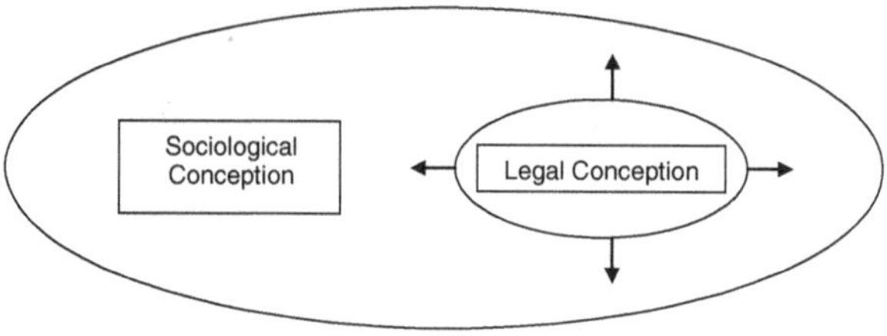

Fig. 1.2 Legal conception of refugees as a subset of the sociological conception

conflict overlap to produce migration (Suhkre, 1994: 488). Suhkre recognises that, in relation to complex displacement, displaced populations share some characteristics with refugees as recognised by international law but argues that:

> the population movement itself reflects different and potentially more devastating causes of displacement than envisaged in classical refugee law and practice.

Suhkre states that the office of the United Nations High Commissioner for Refugees (UNHCR) has a legal obligation to provide protection and relief in situations of complex displacement. This may be so, but when contemporary approaches to the interpretation of the Convention are taken into account, situations of complex displacement may vest not only the UNHCR but also States Parties to the Convention with a legal obligation of protection in some cases. None of this is to deny that the Convention has definitional limits. It *is* narrower in its scope than definitions grounded in a sociological conception of refugeehood. Suhkre's (1994: 488) argument for limiting the use of the term 'refugee' to cases where the central issue is state activity points towards an elemental component of the Convention definition, namely, the failure of state protection. Nevertheless, Suhkre's assertion that persecution 'implies' state conducted, permitted, or encouraged *violence* is mistaken. As discussed below, contemporary approaches to this issue may bring more environmentally displaced persons within the scope of the Convention than those where violence is imminent and without the need for a state of civil war to exist.

As others rightly observe (Black, 2001: 12–13; Castles, 2002: 10; Gibney, 2001: 7), the refugee discourse in Western states has become hijacked by populist concerns which has, in turn, led to more restrictive polices. With increased economic migration and a rise in the numbers of persons claiming refugee status, migration has increasingly been discussed in security terms (Story, 2005: 4; Vogler, 2002: 188). I agree that in developing policy responses to environmentally induced displacement all care must be taken to ensure there is no downgrade in the level of protection afforded to Convention refugees. However, unlike Black and Castles – and here I agree with Renaud et al. (2007: 30) – I believe that there is some value in retaining the concept of an 'environmental refugee'. Unlike the definition Renaud et al. proffer, my conception of the term is narrower, describing only those persons who fall within the existing Convention definition for whom an environmental issue provides a sufficiently relevant context for the presentation of the claim for refugee status.

There are, in my view, three advantages in retaining this conceptual linkage in this narrow context. First, it has an evidential value. As already mentioned, environmental degradation may be the result of policy choices that form part of a wider pattern of state-sanctioned repression. The 'environmental' descriptor is not redundant. Second, it can provide a critical check within Refugee Status Determination (RSD) systems. It may be a difficult task to identify when environmental causes stop and political causes begin (Lee, 1997: 277). Allowing conceptual space for 'environmental refugees' ensures that decision-makers within RSD systems faced with large-scale migration flows do not make a priori assumptions that claimants whose migration occurs against a background of environmental degradation cannot, as a matter of principle, fall within the scope of the Refugee Convention and are to be treated akin to economic migrants. The 'environmental' descriptor also has utility. This is no idle concern. Statements have been made at the highest judicial level that persons 'fleeing natural disasters' cannot obtain Convention-based protection.[5] There is real potential for such statements to be relied on more broadly than warranted.

Third, actual or threatened large-scale migration can provide impetus for radical intervention when the interests of developed states are at stake. Making explicit the environmental component, as opposed to simply highlighting the political component in situations of complex migration, may provide some impetus for developed refugee-receiving states to alter their own policies or encourage change in the policies of international institutions and refugee-producing states, in order to address both drivers.[6]

3 Some Jurisprudential Issues Explored

What follows is a necessarily brief overview of some jurisprudential issues which may arise within two alternative RSD systems – the pan-EU system as set out in the EU Directive on Minimum Standards for Qualification (the Directive) and the New Zealand RSD system. Before doing so, it is important to set out two relevant conceptual underpinnings of the Convention.

3.1 Conceptual Underpinnings

3.1.1 Declaratory in Nature

The Convention and, by extension, domestic RSD processes are declaratory in nature (UNHCR, 1992: 9). In other words, persons are refugees as soon as their

[5] See *A* v *MIMEA* [1998] INLR 1 at p. 19 (High Court of Australia); *AH (Sudan)* v *Secretary of State* [2007] 3 WLR 832 at p. 844 (House of Lords).

[6] See Castles (2002: 10).

circumstances, in fact, meet the Convention's refugee definition – not when those facts are recognised by domestic RSD systems. Essentially, RSD systems function to identify those to whom the receiving state owes the various obligations under Articles 2–34 of the Convention. The principle is recognised under both systems.[7] There is, therefore, no discretionary element. Once the fact-finding inquiry reveals that a person's circumstances meet the Convention's inclusion criteria, the international law duties are owed irrespective of whether the external migration which spawned the application for refugee status takes place against a background of environmental degradation.

3.1.2 Discrimination Driven

At the heart of the Convention is the avoidance of discrimination in the enjoyment of basic human rights (Hathaway, 1991: 112). This is important; the discrimination of marginalised groups by dominant groups is usually a reason for disputes over scarce or degraded resources to escalate (Baechler, 1999: 121, 123). Where the discrimination encountered in the context of environmental degradation interferes with the basic human rights of a marginalised group, this provides a clear avenue by which the circumstances of the marginalised group may fall within the scope of the Convention. This is not to say that they automatically will. The legal condition of 'being persecuted' connotes a particular form of discrimination. Whether their circumstances reach the Convention's threshold will depend on the nature and extent of the discrimination encountered.

3.2 *Jurisprudential Issues*

3.2.1 Alienage

As Renaud et al. (2007: 13) note, alienage – being outside one's country of nationality or former habitual residence – is a key criteria for recognition under the Convention. In many cases, this may be an undesirable step to take or insurmountable hurdle to cross, and many environmentally displaced persons will therefore remain internally displaced. Plainly, however, this will not be so in all cases, and once external migration takes place and a claim is lodged in a receiving state, the issue disappears from RSD process.

3.2.2 Being Persecuted

Under both the Directive and New Zealand systems, 'being persecuted' comprises two essential ingredients: serious harm of some kind and a failure of state protection.[8]

---

[7] See Recital 14 of the Directive and Decisions of the New Zealand Refugee Status Appeals Authority (RSAA) in *Refugee Appeal No. 59* (19 May 1992) and *Refugee Appeal No. 75977* (22 November 2007), pp. 83–85.

[8] See Directive Article 9(1)(a) and *Refugee Appeal Nos. 75228 and* 75229 (19 November 2002), p. 88, respectively.

The question arises as to which human rights count in establishing 'being persecuted'? Where the state response to environmental degradation takes the form of denial of shelter, medical assistance or food aid, an objection might be taken that the type of loss is socio-economic in nature and not, therefore, be the type of harm contemplated by the Convention. Discussing legal conceptions of a refugee, Suhkre (1994: 488) argues that the Western (and hence legal) conception of protection:

> has been understood not as an entitlement to goods, however basic (such as food and shelter), but rather as the absence of fundamental wrongdoing (persecution, for example).

Suhkre goes on to argue that persecution implies that the state itself directs, deliberately permits, or encourages violence towards an individual or group, and that the right to be free of such violence 'constitutes the basis for a refugee's claim to asylum'.

This, however, oversimplifies the position. As Hathaway (1991: 103) observes, the drafters of the Convention never intended a 'monolithic or absolute conceptual standard' to apply. Rather, persecution was always capable of encompassing both basic civil and political freedoms and 'serious social and economic consequences'. In an insightful article, McGregor (1993: 161) notes concern that a narrow 'political' definition of persecution in the Convention may operate to exclude the environmentally displaced. Noting calls for a broader conception based on human rights violations including rights under the 1966 International Covenant on Economic Social and Cultural Rights (ICESCR), McGregor argues that such an approach may make the environmentally displaced 'eligible for international assistance and protection according to whether the suffering amounted to a first order violation of human rights'.

Refugee law has, in a sense, caught up with McGregor. Both RSD systems considered here are capable of embracing a wide range of state activity as part of the inquiry into 'being persecuted'. Both systems use a human rights framework as the yardstick to measure whether someone's circumstances are such as to amount to their 'being persecuted' but utilise different frameworks as their point of reference.

Under the Directive, the relevant framework is the 1950 European Convention on Human Rights and, in particular, rights in respect of which states cannot derogate their obligations, such as the right to life and the prohibition on torture, inhuman or degrading treatment, or punishment. While this definition sets a high threshold, two points are noteworthy. First, it is not the case that only these rights can be relied on. Rather, as Article 9(1)(b) makes clear, 'breaches of other rights may lead to a finding of: being persecuted', provided that any harm arising from the breach reaches the level of seriousness contemplated by the non-derogable rights. Second, Article 9(2)(b) recognises that the discriminatory implementation of administrative measures may be a form of persecution. The discriminatory denial[9] of food, shelter, and medicine by state administrative bodies is plainly capable of falling within the scope of these articles.

By contrast, New Zealand's refugee law looks primarily to the triumvirate of international law treaties known as the International Bill of Rights. The Refugee Status Appeals Authority (RSAA) has long adopted a human rights framework – most recently and fully articulated in *Refugee Appeal No. 74665* (7 July 2004) at [56]–[80]. In *Refugee Appeal Nos. 75221 and 75225* (23 September 2005) at [79]–[112], the Authority recognised that socio-economic rights under the ICESCR properly inform the inquiry as to 'being persecuted'. It has recognised that in situations where food and other essential survival items are being withheld on a discriminatory basis; so that an individual faced a real risk of starvation, the right to adequate food, coupled with the right to life can bring this situation within the scope of 'being persecuted' for the purposes of the Convention – see *Refugee Appeal No. 74665* at [89].

The crucial point to stress is that jurisprudence relating to 'being persecuted' *is broader than is assumed in much of the literature to date*. Conceptually, 'being persecuted' is not limited to issues of 'political violence'. While difficult evidential issues may arise, the concept is nevertheless sufficiently flexible to include the discriminatory denial of otherwise available basic social goods, where this would result in serious harm of an acute kind.

3.2.3 Internal Protection Alternatives (IPA)

A feature of both RSD systems is that refugee status may potentially be denied where there is a part of the country of origin in which there is no risk of the claimant being persecuted. This will prove to be a significant issue in cases where the underlying environmental issue is localised. Both systems require any site of internal protection to be free of persecution or other forms of serious harm – see Directive Article 8 and *Refugee Appeal No. 71684/99* (29 October 1999) at [55]–[62]. In cases of localised environmental harm, it will be critical to fully consider the nature of the state in question. For example, in a state with a strong security-intelligence apparatus and poor human rights record, a person who complains about discriminatory state activity in relation to a situation of localised environmental degradation may well be imbued with a state-wide negative political profile and face a real risk of being persecuted on a state-wide basis. If so, a viable IPA may well not exist – see *Refugee Appeal No. 71684/99* at [55]; *Januzi* v *Secretary of State for Home Department* [2006] 2 AC 426 at [21] and [48] (House of Lords).

[9] This discriminatory denial serves to distinguish these types of cases from those where the receiving state is simply unable to provide essential health care. Recent decisions of the European Court of Human Rights have increased the threshold for obtaining relief against expulsion in these cases: compare *D* v *UK* [1997] 24 EHRR 423 with *N* v *UK* [2008] Imm. AR 657.

Article 8 additionally requires European RSD systems to also enquire whether it is, in all the circumstances, 'reasonable' to expect the claimant to stay in the site of IPA. In contrast, the New Zealand system has sought to quantify this by reference to the standard of protection afforded to refugees under the Convention itself – see *Butler* v *Attorney General* [1999] NZAR 205 at p. 218, *Refugee Appeal No. 71684/99* at [57] and *Refugee Appeal No. 76044* (11 September 2008) at [132]–[139]. This provides for the provision of basic norms of civil and political as well as economic and social rights. In practice, it is open to debate whether this difference will have a significant impact in cases of environmental displacement – after all, it can hardly be reasonable to expect an environmentally displaced person to stay in the proposed site of an IPA where these basic norms are not provided for.

3.2.4 Nexus to a Specified Convention Ground

As a core component of the Convention, the requirement of a causal connection between the state of 'being persecuted' and a protected Convention ground is naturally reflected in both systems – see Directive Article 9(3) and *Refugee Appeal No. 71427* (16 August 2000) at [111]. Here, issues of mixed motivation behind migration and the complexity of linkages between environmental degradation, conflict, and migration may become important. Any environmentally displaced person wishing to access the Convention's protection regime must show that their situation not only reaches the threshold of being persecuted, but also that this is *by reason* of their race, religion, nationality, membership of a particular social group, or their political opinions. In situations where the state, without discrimination, is simply unable to respond to natural disaster, this will prove fatal to any claim for recognition under the Convention, even if the claimant faces a situation of serious harm. It is this category of environmentally displaced persons for whom the need for an alternative protection regime is most acute. Detailed and accurate research into state responsibility for both the causes and consequences of environmental degradation will therefore be vital to assist in establishing whether the required nexus exists.

As to the standard by which the nexus must be considered, in New Zealand, all that need be established is that the Convention ground contributes to the risk of being persecuted – see *Refugee Appeal No. 72635* (6 September 2002) at [173]. The Directive is silent on the issue. However, it largely reflects English law which has yet to settle on a definitive standard. Presented usually as a choice between a 'but for' test and an 'effective cause' test, English courts have rather left the issue to be resolved by applying common sense to the facts – see *Shah and Islam* v *SSHD* [1999] Imm. L.R. 283 (House of Lords). That said, aspects of the House of Lords judgement in *R* v *SSHD, ex p Sivakumar* [2003] UKHL 14 at [30] suggest that the Convention ground need only partially contribute to the persecutory act.

4 Conclusion

As the debate progresses as to what the institutional and normative response of the international community should be to environmentally induced displacement, it is important that we do not lose sight of what responses the current international law framework may provide. In particular, it is important that the possible protection afforded by existing international human rights, humanitarian, and refugee law is not overlooked. In this context, this chapter has argued that caution needs be exercised before rigid and immutable distinctions are drawn between environmentally displaced persons and those to whom the Convention's protection regime can extend. While plainly not capable of providing protection to all whose displacement is linked to environmental degradation or resource scarcity, under certain conditions which are by no means uncommon, the Convention's definition of 'refugee' may extend protection to some.

If projections by the IPCC (2007: 13) as to the numbers of persons potentially affected by climatic change in the coming decades prove accurate, the issue is only going to become more prominent. While it would be a mistake to assume that climate change will inevitably lead to widespread conflict or that all or even most people adversely affected by climate change will respond (adapt) by migrating externally (Piguet, 2008: 6–7), it would only take a small percentage of these persons affected to do so for RSD systems, where they exist at all, to be placed under stress.

This suggests that a number of responses are required. First, as Castles (2002: 7) observes, an increased sensitivity to environmental factors in research into complex migration may result in a greater number of migration flows being seen as having a significant environmental component. While important for making the case for the need for an alternative protection regime, this research may also provide vital evidence to support a claim under the Convention. Detailed and good quality research is, therefore, essential. Second, not all states which may receive environmentally displaced persons (e.g. India and Malaysia) are signatories to the Convention. There should be a push for them to become so. Third, many environmentally displaced persons may be forced to migrate to States Parties to the Convention with underdeveloped RSD systems (Gibney, 2004: 11; Hathaway, 2007: 92; Kelley, 2007: 407; UNHCR, 2007: 27) which are ill-equipped to cope with the complexity of the issues raised. There is a need for capacity building, so that those environmentally displaced persons in such states, whose circumstances mean they are in fact Convention refugees, get the recognition and protection they deserve and are entitled to.

References

Ashton, P.J. (2002): Avoiding conflicts over Africa's water resources. *Ambio* 13(2), 236–242.

Baechler, G. (1999): Environmental degradation in the south as a cause of armed conflict. In: Carius, A., and Lietzmann, K.M., (eds.), *Environmental Change and Security*. Springer, Berlin. pp. 107–129.

Baird, R. (2008): Climate change and minorities. In: *The State of the World's Minorities*. Minority Rights Group International, London. pp. 9–19.

Barnett, J. (2003): Security and climate change. *Global Environmental Change* 13, 7–17.

Betts, A. (2005): International cooperation between North and South to enhance refugee protection in regions of origin. *Working Paper Series*. Working Paper No. 25. Refugee Studies Centre, University of Oxford.

Black, R. (2001): Environmental refugees: myth or reality? *New Issues in Refugee Research*. Working Paper No. 34. United Nations High Commissioner for Refugees (UNHCR), Geneva.

Castles, S. (2002): Environmental change and forced migration: making sense of the debate. *New Issues in Refugee Research*. Working Paper No. 70. United Nations High Commissioner for Refugees (UNHCR), Geneva.

Conisbee, M., and A. Simms. (2003): *Environmental Refugees: The Case for Recognition*. New Economics Foundation, London.

El-Hinnawi, E. (1985): *Environmental Refugees*. United Nations Environment Programme, Nairobi.

Gibney, M.J. (2001): The state of asylum: democratization, judicialisation and evolution of refugee policy in Europe. *New Issues in Refugee Research*. Working Paper No. 50. United Nations High Commissioner for Refugees (UNHCR), Geneva.

Gibney, M.J. (2004): *The Ethics and Politics of Asylum*. Cambridge University Press, Cambridge, MA.

Gleditsch, N.P. (1998): Armed conflict and the environment. *Journal of Peace Research* 35(3), 381–400.

Harbom, L., and P. Wallensteen. (2005): Armed conflict and its international dimensions. *Journal of Peace Research* 42(5), 623–635.

Hathaway, J. (1991): *The Law of Refugee Status*. Butterworths, Toronto, ON.

Hathaway, J. (2007): Why refugee law still matters. *Melbourne Journal of International Law* 8, 89–103.

Homer-Dixon, T., and V. Percival. (1996): *Environmental Scarcity and Violent Conflict: Briefing Book*. University of Toronto Project of Environment, Population and Security, Toronto, ON.

Hugo, G. (1996): Environmental concerns and international migration. *International Migration Review* 30(1), Special Issue Ethics, Migration and Global Stewardship, 105–131.

HRW. (2003): *The Iraqi Government Assault on the Marsh Arabs*. Briefing Paper, New York, NY.

HRW. (2009): *World Report 2009 – Events of 2008: Burma*, New York, NY.

IPCC. (2007): *Climate Change 2007: Impacts, Adaptation and Vulnerability. Summary for Policymakers Contribution of Working Group II to the Fourth Assessment Report of the Intergovernmental Panel on Climate Change*. IPCC, Brussels.

Jacobsen, J. (1988): *Environmental Refugees: A Yardstick of Habitability*. World Watch Paper No. 86, World Watch Institute, Washington, DC.

Kelley, N. (2007): International refugee protection challenges and opportunities. *International Journal of Refugee Law* Oxford University Press, Oxford, 401–439.

Lee, S. (1997): In limbo: Environmental refugees in the third world. In: Gleditsch. N., (ed.), *Conflict and the Environment*. Kluwer Academic Publishers, Dordrecht. pp. 273–292.

Levy, M.A. (1995): Is the environment a national security issue? *International Security* 20(2), 35–62.

Lonergan, S., and A. Swain (1999): *Environmental Degradation and Population Displacement.* AVISO Bulletin No. 2 Global Environmental Change and Human Security Project University of Victoria, Canada. Available at http://www.gechs.org/aviso/o2/index.html

McGregor, J. (1993): Refugees and the environment. In: Black, R., and Robinson, V. (eds.), *Geography and Refugees: Patterns and Processes of Change.* Belhaven Publishers, New York, NY. pp. 157–170.

Olson, M.E. (1979): Refugees as a special case of population redistribution. In: Gosling, L., and Lim, L., *Population Redistribution: Patterns, Policies and Prospects.* New York: United Nations Fund for Population Activity. pp. 130–152.

Pelling, M., and K. Dill. (2006): 'Natural' disasters as catalysts of political action. In: *Human Security and Resilience* ISP/NSC Briefing Paper 06/01. Royal Institute of International Affairs, London.

Piguet, E. (2008): Climate change and forced migration. *New Issues in Refugee Research.* Working Paper No. 153. United Nations High Commissioner for Refugees (UNHCR), Geneva.

Raleigh, C., L. Jordan, and S. Salehyan (2008): *Assessing the Impact of Climate Change on Migration and Conflict.* Paper Commissioned for Social Dimensions of Climate Change Workshop, Social Development Department, World Bank, Washington, DC.

Renaud, F., J. Bogardi, O. Dun, and K. Warner (2007): Control adapt or flee: How to face environmental migration? *Intersections* No. 5. UNU-EHS, Bonn.

Richmond, A. (1993): The environment and refugees: Theoretical and policy issues. *Revised version of paper presented at meetings of the International Union for the Scientific Study of Population,* Montreal, Canada.

Ronnfelt, C.F. (1997): Three generations of environment and security research. *Journal of Peace Research* 34(4), 473–482.

Story, B. (2005): *Politics as usual: the criminalization of asylum seekers in the United States.* Working Paper No. 26. Refugee Studies Centre, Oxford University, Oxford.

Suhkre, A. (1994): Environmental degradation and population flows. *Journal of International Affairs* 47(2), 473–496.

Suhkre, A. (1997): Environmental degradation, migration, and conflict. In: Gleditsch. N., (ed.), *Conflict and the Environment.* Kluwer Academic, Dordrecht. pp. 255–272.

Turton, D. (2003): *Refugees and 'Other Forced Migrants'.* Working Paper No. 13. Refugee Studies Centre, Oxford University, Oxford.

Vlachos, E. (1997): Environmental refugees: The growing challenge. In: Gleditsch, N., (ed.), *Conflict and the Environment.* Kluwer Academic Publishers, Dordrecht. pp. 293–312.

Vogler, J. (2002): The European Union and the 'securitisation' of the environment. In: Page, E.A., and Redclift, M. (eds.), *Human Security and the Environment.* Edward Elgar, Cheltenham. pp. 179–198.

UNHCR (1992): *Handbook on Procedures and Criteria for Determining Refugee Status Under the 1951 Convention and the 1967 Protocol Relating to the Status of Refugees,* Geneva.

UNHCR (2007): *Statistical Yearbook 2006: Trends in Displacement, Protection and Solutions,* Geneva.

Chapter 2
Institutional Barriers to the Recognition and Assistance of Environmentally Forced Migrants

Chloe Anne Vlassopoulos

1 Introduction

The recent increase in the number of publications, debates, and public discourse about the displacement of people caused by environmental changes can be reasonably attributed to the alarming intensity of the situation. In the near future, the proliferation of environmentally forced migrants (EM) is expected to create tremendous socio-economic and political problems. In spite of the immensity of the problem, no official definition of this class of migrants has been established and no policy measures are adopted at the national and international level.[1]

The absence of any real effort to define the status of environmental migrants is often linked to the deliberate effort not to include them in the class of (political) refugees and consequently grant them the protection provided by the Geneva Convention (Kibreab, 1997). Thus, the effort of the advocates of an ad hoc policy for the EM amounts to persuading the political community that those migrants should also be considered as persecuted (Conisbee and Simms, 2003; Myers, 1997). This reasoning has found support in the prevailing political tendency to limit immigration flows.

Another explanation given to this 'agenda denial' (Cobb and Ross, 1997) is related to the uncertainty of the phenomenon. All figures advanced on the displacement of millions of people cover periods as distant as 2025 or 2050 (Myers, 2002). Many scholars remain sceptical and contest those numbers (Black, 2001, Castles, 2002). They emphasise the absence of certainty about these apocalyptic scenarios and ask for more analysis of the complex factors behind them. Governments hesitate to engage in policy action in a context of uncertain long-term impacts. In response to this situation, more case studies are requested in order to prove the necessity for further consideration.

C.A. Vlassopoulos (✉)
CURAPP-CNRS/University of Picardie Amiens, France
e-mail: chloe.vlassopoulou@u-picardie.fr

[1] In order not to create confusion with traditional refugees, I use the terms environmental (forced) migrant and environmentally displaced persons.

T. Afifi, J. Jäger (eds.), *Environment, Forced Migration and Social Vulnerability*,
DOI 10.1007/978-3-642-12416-7_2, © Springer-Verlag Berlin Heidelberg 2010

However, the advocates of the asylum regime and the uncertainty of the future are not the only reasons for the difficulties in this agenda-setting process. A further explanation of failure appears to be much more complex and largely related to the specificity of the issue which involves different policy sectors. Each one of these sectors has its own rationality and its policy goals and priorities. Therefore, a compromise on problem definition and policy content becomes particularly difficult.[2]

The first part of this contribution seeks to present the complex process of public problem definition. The second part proposes an analysis of the constraints imposed by the multi-sectoral aspect of the issue. The third part discusses the recent evolution of the international debate that restrictively defines the environmental migrant as a climate migrant. We suggest that this effort to define the identity of those who need protection paradoxically leads to postponement of any attempt to officially recognise environmentally displaced persons as an autonomous public problem and to propose specific policy measures. The final section breaks with the globalising and the restrictive definitional strategies outlined in the previous parts. It suggests a more pragmatic approach consisting of the use of present policy instruments in order to give different policy sectors the possibility to help environmentally vulnerable populations. This could be a first stage for further negotiations for a more integrated approach in the future.

2 The Russian Dolls or the Definition of Public Problems

Public problems are like Russian dolls that are enclosed within one another. Let us take the example of the environmental migrants: Greenhouse gas (GHG) emissions cause global warming. Global warming is a potential force for creation and/or expansion of deserts. This subsequently leads to loss of biodiversity, famine, and forced migration. Migration generated by desertification can be defined as an autonomous public problem with further consequences like conflict, cultural, and economic impoverishment in the departure regions, human, and ecological vulnerability, etc. But it can also be perceived as a consequence of desertification.

Greenhouse gases → Global warming → Desertification → Migration → Conflict, etc.

As Edelman (1991) suggests, very few situations are recognised as a public problem and get registered on the political agenda. The recognition of a problem does not occur automatically; it depends on various parameters not necessarily related to the gravity of the situation. Serious problems can remain

[2] This chapter is based on observations and personal interviews with agents of the French and Indian administrations and the United Nations (Geneva). I wish to thank them for having accepted to meet me and respond to my questions.

far from the attention of governments, while others, perhaps less pressing, can be the subject of courageous measures. The official recognition of a problem means allocation of a budget, distribution of competences, and designation of persons to blame and protect. The definition (or not) of a situation as a problem and its official registration (or not) on the agenda are purely political actions that involve the search for a compromise between multiple actors supporting different positions. To put it another way, problem definition is an issue of power, since it divides the social and public actors into winners (who profit from the recognition of the problem) and losers (who will lose material or symbolic resources) (Stone, 1997). For example, the recognition of environmental migrants as an autonomous problem will strengthen the positions of the agency asked to manage this problem, which will probably in turn have its budget and its personnel increased. It will also help people affected by the deterioration of the environment who will benefit from positive measures. On the other hand, this same recognition can prove harmful for the migrants. It may also meet the dissatisfaction of the polluters.

When a situation becomes a public problem, it passes through a double definitional debate (Vlassopoulos, 2007). The first determines the causes of the problem and answers the question 'what is the problem?'. The second determines the consequences of the problem and answers the question 'why does this situation constitute a problem?'. The causes identify the measures to be adopted and the agents who have competence to implement them. Each reference to the consequences of the problem constitutes not only a justification legitimising to differing degrees of the intervention of political leaders ('we act to protect the citizens' health', 'we act to guarantee safety', etc.), but also a means for the recognition of the authority of those who claim to have one or another type of competence.

The multi-sectoral character of the issue of environmental migrants disturbs this double definitional process preventing its definition as an autonomous public problem.

3 The Difficulty of Defining a Multi-sectoral Problem

Since the El-Hinnawi (1985) report, the attempts to define environmental migrants proved in fact that this problem is situated at the intersection of different policy sectors, each one with its own rationality, objectives, and functioning. The environmental causes of the problem require the involvement of the environmental sector, both at the national and international level. The consequences of the problem are not, however, only environmental; according to the dominant discourse, they relate to the humanitarian sector (human suffering) or to the security sector (risk for conflict). In other words, the gravity of the problem is not justified with reference to ecological crisis but with reference to the human conditions of living and survival (Fig. 2.1).

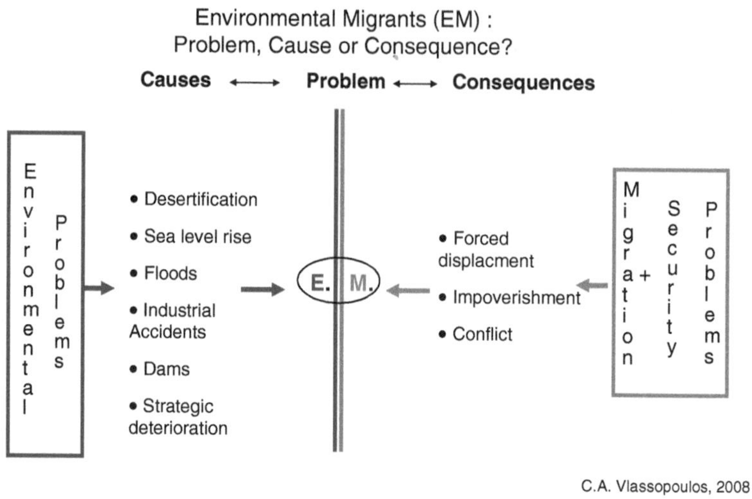

Fig. 2.1 Environmental migrants: problem, cause, or consequence

The environmental policy network comprises actors traditionally in charge of combating environmental deterioration (United Nations Environmental Programme (UNEP), environmental NGOs, Departments of the Environment, etc.). It is not by accident that the term 'environmental refugee' emerged for the first time on the international arena in the El-Hinnawi report published by UNEP. The policy network competent on the humanitarian and security issues comprises very different actors with no environmental competences (United Nations High Commissioner for Refugees (UNHCR), United Nations Development Programme (UNDP), Office for the Coordination of Humanitarian Affairs (OCHA), International Organizations for Migration (IOM), Ministries of Interior and Foreign Affairs, humanitarian NGOs, etc.).

Important incompatibilities separate these groups of actors, making their collaboration difficult. More specifically, the environmental sector was built in the 1970s on an ecocentric approach to reality against the anthropocentric approach dominant until then (Theys, 2005). The environment constitutes a new autonomous social value that must be protected in itself and not, as in the past, through reference to human health. The new environmental policies are based on the responsibility concept operationalised by the 'polluter-pays' principle. They also refer to the prevention and precautionary principles which express the need for decision-makers to anticipate harm to the public or to the environment.

The humanitarian as well as the security sectors are built on an anthropocentric approach to reality. This involves giving priority to human well-being. Humanitarian and security policies are urgent responses to human crisis situations, and as such, they cannot be based on the individual responsibility concept. Costs are shared by the community.

From what precedes, it becomes apparent that very different public policies can be adopted in order to deal with EM according to the policy sectors involved. The environmental sector concentrates primarily on the causes of the problem. It proposes either ex ante policy measures in order to prevent environmental degradation or ex post restoration measures, asking the polluters to internalise the cost of pollution in order to mitigate environmental degradation. The humanitarian sector focuses on the consequences of the problem by proposing ex post measures of relief through the collectivisation of the induced costs. The security sector concentrates on the prevention and/or regulation of migration flows.

Neither the objectives (environmental protection vs. human protection) nor the means (constraint and sanction, help, and inducement) are common to these sectors of public action. Stone (1997) rightly affirms that a problem is never defined on the public agenda in all its dimensions. It always passes through a process of simplification that reduces the scope of the problem and makes it more manageable. This is occurring in recent attempts to restrictively redefine environmental migrants as climate migrants. This redefinition focuses attention on the biggest sub-category of environmentally displaced people, but it runs the risk of diluting the problem within the very large issue of climate change and making the discourse on migration much less audible.

4 From Environmental to Climate Migrants

The different numbers of environmental migrants and the approaches and sectors involved push the actors to seek clarification of the term. This evolves through a process of inclusion and exclusion which finally defines both the content of policy, i.e. those to be protected by policy measures and at the same time those who are competent to make policy. Different arenas of discussion mostly point to climate migrants by putting aside all other cases of environmental displacement (Felli, 2008). We notice that European Greens concentrate the debate exclusively on climate migrants.[3] This is also the case within various UN agencies.

I suggest that this process in progress takes the form of a 'double fault' of agenda setting. On the one hand, it excludes from the public debate all other categories of population suffering from environmental crises that cannot hope any more for protection and compensation (industrial accidents, dam construction, war, . . .). On the other hand, the good position that climate change has on the political agenda does not seem to benefit the environmental migrants as the advocates of this alignment would have hoped. Paradoxically, the annexation of the problem to the climate change issue tends to slow down the effort to consider environmental migration as an autonomous public problem.

---

[3] See Conference on Climate Refugees at the European Parliament, the 11 June 2008 and subsequent Declaration.

Pushed by the reports of the Intergovernmental Panel on Climate Change (IPCC), climate change is imposed on the national and international agendas as a major public concern. Important human and financial resources are engaged both for research and policy making, but the way this issue is being defined does not leave any place for discussion on migration.

Historically, since 1992 the policy to combat the climate change has passed from a traditional environmental approach to a development-oriented approach. The United Nations Framework Convention on Climate Change (UNFCCC) and the Kyoto Protocol provide binding agreements for the reduction of GHG emissions (mitigation strategy). However, this environmental approach has so far proved to be insufficient not only due to the lack of engagement by certain big GHG-emitter countries but also in relation to the inappropriate application of the emissions trading system. Therefore the discourse on mitigation has started to weaken giving place to another policy-oriented discourse concentrating on adaptation strategies.

The decade following the first IPCC Assessment report saw significant evolution of climate change impacts and adaptation research, due to a number of factors (Smit et al., 2000). As pointed out in the report of the Canadian Climate Change Impacts and Adaptation Program, there was growing evidence that climate change was already occurring. Impacts could no longer be seen as hypothetical outcomes but instead needed to be addressed as imminent concerns. Further, it had become apparent that mitigation could not prevent climate change from occurring even if stabilisation of carbon dioxide were achieved. Thus the international climate change community considered adaptation as a principal policy objective. This shift in attitude is reflected in the changing titles of the three IPCC Working Group II assessment reports completed between 1990 and 2001: 1990: *Impacts Assessment of Climate Change*; 1995: *Impacts, Adaptation and Mitigation of Climate Change*; 2001 and 2007: *Impacts, Adaptation and Vulnerability*.[4] The emphasis is consequently put on the most vulnerable regions, which generally coincide with the poorest regions lacking the necessary infrastructures to cope with the effects of global warming (droughts, floods, extreme weather events). In other words, the environmental policies no longer appear as the principal response to the problem. Development projects offering improved resilience to vulnerable regions become the principal instrument for climate policy today.

In that context, the main policy actors are to be found in the development sector and not in the environmental and humanitarian sectors. As indicated by humanitarian agents, 'the logic of development policies does not meet the logic of humanitarian policies'. This is not the case for the environment which – since the Rio Conference – has no longer been seen as an enemy but as a partner for (sustainable) development. The concept of sustainable development reflects, however, the integration of the environment into the need for development and

[4] http://www.adaptation.nrcan.gc.ca.

not vice versa.[5] This evolution towards development-oriented policies which put the emphasis on adaptation to deal with climate change slows down the efforts to define environmental migration as a new global public issue. In fact, adaptation strategies are in contradiction with the recognition of this issue: taking into account environmental migrants means anticipating the insufficiency of adaptation efforts. In other words, in a context where adaptation becomes the key strategy for climate policy, environmental migrants become a non-issue.[6]

5 Towards a More Pragmatic Approach to the Problem?

According to the discussion above, two different strategies have been developed in the last 20 years by the advocates of people who need to leave their traditional homelands because of an environmental disruption. The first consists of seeking a broad definition including all kinds of environmental causes of displacement. The second strategy tends to promote a restrictive definition of 'climate' migrants. In both cases, the shift from discourse to action has not succeeded for multiple reasons, leaving people suffering from environmental problems without systematic assistance.

The effort to impose a broad definition meets with an unfavourable political and economic context, pushing European states to close their borders and restrain the migration flows. Paradoxically, the alarmist discourse on the future millions of environmental migrants does not stimulate a call for immediate action but rather strengthens the tendency to control migration. At the same time, the multiple figures of environmental migrants mean that the public is unclear about who the environmental migrants really are. The restrictive definition, as mentioned above, leads to the incorporation of the migration issue into climate change policy. This constitutes to some extent an ex-ante problem-solving inasmuch as the dominant approach consists of helping the regions to adapt to change, so that people do not see themselves forced to flee their homelands. Although in this case the migrants' portrait appears to be clearer and more concrete, the alignment between migration and climate change tends to exclude the former issue from the political agenda and to strengthen the importance of the latter.

Renaud et al. (2007) rightly affirm that 'scientific "concerns" instead of the pragmatic application of a precautionary principle paralyse both the scientific and the policy making communities'. They propose then to distinguish between different sub-classes of environment-related displacements in order to better

[5] We do not enter here into the debate relating to the no-growth movement as formulated by Georgescu-Roegen (1979) or more recently by Christophe (2007). I want only to emphasise the fact that the notion of sustainable development imposes an implicit hierarchy between development and the environment.

[6] Personal interviews allowed us to note the non-consideration of the question of migration during the discussions and research on adaptation strategies.

understand the urgency of the assistance that is needed. The criterion of classification is in this case the nature of the environmental degradation. The concern for a pragmatic approach to the problem needs also to take into account the compatibility and rationality of the sectors addressing the policy to be adopted. Therefore, we propose to separate the problem into sub-issues corresponding to the specificities of the different relevant sectors.

The strategies proposed until now concentrate on the most conflicting dimension of the policy process: the problem definition. They ask, as a first step, to officially recognise a specific status offering protection and compensation to EM. This eminently political process is accompanied by strong media coverage, and political and institutional conflicts paralyse negotiations for future policy making. The fragmentation of the problem in sub-issues divides the negotiation framework into smaller and more integrated policy communities that can facilitate the debate and the adoption of concrete measures. This does not involve an overall agreement but rather multiple compromises within the various sectors concerned. These compromises may go in the direction of adjustments to existing policy instruments. In this way, they could also respond to the needs of environmental migrants.

A sector-relevant criterion to carry out the division of the problem is the possibility or not to identify those to blame for environmental disruption and allocate the compensation costs to them. The application of this criterion to the environmental causes of migration listed by Lonergan (1998) distributes the different environmental crises to three more or less separate arenas of negotiation.

Figure 2.2 presents three cases of responsibility allocation: the case where the actors behind environmental disruption are more or less identifiable and forced

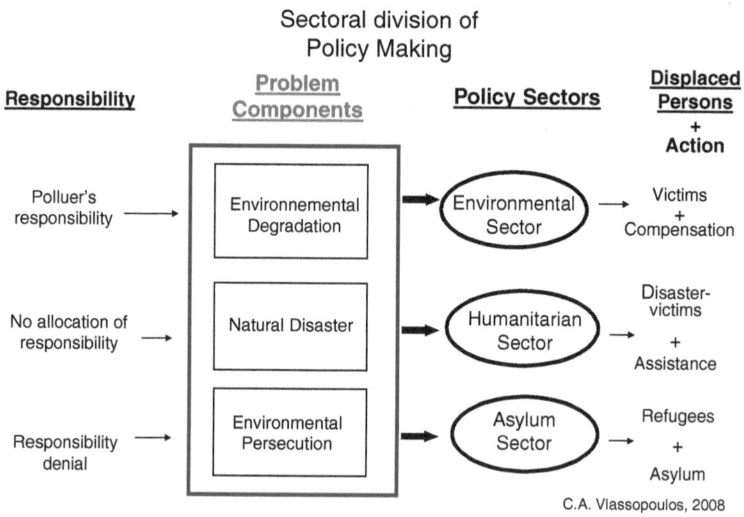

Fig. 2.2 Sectoral division of policy making

to assume the costs of the damage; the case where the attribution of responsibility is impossible; and finally, the case where actors are identifiable, but their responsibility is not assumed. In each one of these cases, the response mechanism to displaced populations differs, the qualification of people affected is different, and the measures to engage are different (Vlassopoulos, 2008).

In the first case (environmental degradation), polluters are recognised, and the official authorities can organise a compensation system.[7] This case includes many of the causes of environmental displacement:

- *Desertification* is a problem above all attributed to human activity (overexploitation of the land, persistence of inadequate farming practices, ...) and worsened by certain natural phenomena like long periods of drought – partly also related to human behaviour.
- *Deforestation* is also a problem involving human behaviour (overexploitation of the forest resources, transformation of wooded land into agricultural fields, voluntary fires).
- *Climate change* is attributed today mainly to the human activity.
- *Industrial accidents* mostly have quite clearly identifiable polluters.

All of these problems have generated international agreements (Convention to Combat Desertification, Convention on Biodiversity, Climate Convention, Convention on the Transboundary Effects of Industrial Accidents, ...). Those agreements identify the victims as well as the polluters who have to assume the costs of the environmental damage. In that sense, they can be mobilised to respond to the needs of environmental migrants. This will require, however, a shift towards a less ecocentric vision of the reality, recognising human rights and dignity (not only human health stricto sensu) as objects protected by the environmental rules.

In the second case (natural disasters), there is no allocation of responsibility, because environmental disruption and migration are the result of natural events. The question here is where to locate the boundary line between natural origin and anthropogenic origin. Only earthquakes and volcanic eruptions seem clearly to be natural disasters. An answer can only come from a high-level political discussion which will unavoidably simplify reality. This distinction is necessary not only for the clarification of competences between agencies but also for the share of financing between a 'polluter compensation fund' and a 'humanitarian relief fund'.

The third case (environmental persecution) concerns responsibility denial on behalf of the drivers of the environmental disruption (most of the time political authorities) and the inability of the international community to oblige them to assist the victims. This case includes *dam construction* which generates mass population displacements not always protected by their authorities. It also

[7] Bates (2002) proposes to differentiate between polluting activities that intentionally produce migration and activities that unintentionally produce migration. This distinction risks introducing a soft framework within which polluters could easily escape from their responsibilities.

includes *strategic deterioration of the environment* during armed conflict. In both cases, it is possible to consider that the people affected cannot count on the assistance of their governments, and thus, they can be considered as being persecuted. Therefore, the asylum sector needs to engage in discussions about the way to reinterpret the Geneva Convention in order to give these populations 'refugee' status. This will require a shift towards a less anthropocentric vision of the reality, recognising the environment as a cause of human suffering.[8]

In some extreme cases like the disappearance of a country caused by a serious and irremediable deterioration of its environment, the same mechanisms could be mobilised. If the convention on apatrids (i.e. persons unable to certify their nationality) obtains greater recognition within the international community, it could also serve as a means to assist those people. In that direction, the report of the Executive Committee for the programme of the High Commissioner on the apatrids in 2006 encouraged better mobilisation for the identification, prevention, reduction, and protection of stateless persons.

6 Conclusion

This contribution contests the thesis which claims that the environment is not a real cause of migration, and therefore, specific protection is not justified for people experiencing environmental stress. The environmental deterioration causes human and environmental suffering. Prevention as well as assistance and compensation are needed, regardless of the number of persons concerned. The decision to impose a specific status for environmental migrants would probably be the best solution. But both the socio-economic context and the multi-dimensional content of the issue seriously decrease the feasibility of this solution. Restrictively pointing to climate migrants not only excludes from the public debate all other categories of population suffering from environmental deterioration but also, paradoxically, slows down the effort to consider environmental migration as an autonomous public problem.

The response to inaction could be a more fragmented policy-making process that fits better into the traditional institutional divisions and rationalities. Dividing the issue into different more homogenous arenas of negotiation with specific competences can push forward the debate and lead to compromises. Giving clear answers to complex problems is not a feasible target. Simplification is always necessary in the public sphere in order for the agenda-setting process and policy formation to succeed.

---

[8] The recent article by the High Commissioner for Refugees, A. Guterres, goes in that direction by recognizing that 'some movements likely to be prompted by climate change could indeed fall within the traditional refugee law framework' (Guterres, 2008).

References

Bates, D. (2002): Environmental refugees? Classifying human migration causes by environmental change. *Population and Environment* 23(5), 465–477.

Black, R. (2001): Environmental refugees: Myth or reality? *New Issues in Refugee Research.* Working Paper No. 34.UNHCR, Geneva.

Castles, S. (2002): Environmental change and forced migration: Making sense of the debate. *New Issues in Refugee Research.* Working Paper No. 70. UNHCR, Geneva.

Christophe, B. (2007): *L'entreprise et la décroissance soutenable. Réussir les reconversions écologiques.* L'Harmattan, Paris.

Cobb, R.W., and M.H. Ross (eds.) (1997): *Cultural Strategies of Agenda Denial? Avoidance, Attack and Redefinition.* University Press of Kansas, New York, NY.

Conisbee, M., and A. Simms. (2003): *Environmental Refugees. The Case for Recognition.* NEF Pocketbook, London.

Edelman, M. (1991): *Pièces et règles du jeu politique.* Seuil, Paris.

El-Hinnawi, E. (1985): *Environmental Refugees.* UNEP, Nairobi.

Felli, R. (2008): Justice globale pour les réfugié·e·s climatiques? Vlassopoulos, C.A. (dir.), Exodes écologiques: l'environnement source de migrations. *Asylon(s)*, n 6, nov. 2008. http://www.reseau-terra.eu/rubrique154.html

Georgescu-Roegen, N. (1979): *Entropie-ecologie-Economie.* Sang de la Terre, Paris.

Guterres, A. (2008): *Climate change, natural disasters and human displacement. A UNHCR perspective.* UNHCR Paper.

IPCC. (1990): *Impact Assessments of Climate Change,* Working Group II, Australian Government Publishing Service Marketing Section.

IPCC. (1995): *Impacts, Adaptation and Mitigation of Climate Change*, Working Group II, Cambridge University Press, Cambridge.

IPCC. (2001): *Climate Change 2001: Impacts, Adaptation and Vulnerability.* Working Group II, Cambridge University Press, Cambridge.

IPCC. (2007): *Impacts, Adaptation and Vulnerability.* Working Group II, Cambridge University Press, Cambridge.

Kibreab, G. (1997): Environmental causes and impact of refugee movements: A critique of the current debate. *Disasters* 21(1), 20–38.

Lonergan, S. (1998): The role of environmental degradation in population displacement. *Environmental Change and Security Project Report*, Issue 4 (Spring), 5–15.

Myers, N. (1997): Environmental refugees. *Population and Environment* 19(2), 167–182.

Myers, N. (2002): Environmental refugees: A growing phenomenon of the 21st century. *Philosophical Transactions of the Royal Society of London: Series B* 357, 609–613.

Renaud, F., J.J. Bogardi, O. Dun, and K. Warner. (2007): Control, adapt or flee. How to face environmental Migration? *Intersections*, UNU-EHS, no. 5.

Smit, B., I. Burton, R. Klein, and J. Wandel. (2000): An anatomy of adaptation to climate change and variability. *Climatic Change* 45(1), 223–251.

Stone, D. (1997): *Policy Paradox. The Art of Political Decision Making.* W.W. Norton, New York, NY, London.

Theys, J. (2005): Pourquoi les préoccupations sociale et environnementale s'ignorent mutuellement. Un essaie d'interprétation à partir du thème des inégalités écologiques. In: Cornu P. (2007) *Environnement et inégalités sociales*, Ed. Université Libre de Bruxelles.

Vlassopoulos, C.A. (2007): "Double speed policy change". A postpositivist framework for the study of clean air policy in France and Greece. ECPR Paper, Helsinki. http://www.reseau-terra.eu/IMG/pdf/Vlassopoulou.ECPR.pdf

Vlassopoulos, C.A. (2008): Exodes écologiques et réfugiés environnementaux. Quel problème? Quelles approches? In: *Plein Droit,* no. 58. http://www.reseau-terra.eu/IMG/pdf/Vlassopoulou.GISTI.pdf

Chapter 3
What's in a Name: Social Vulnerabilities and the Refugee Controversy in the Wake of Hurricane Katrina

François Gemenne

1 Introduction: A Social Disaster in New Orleans

Abundant literature on disasters has shown that a natural hazard does not always result in a disaster. It only does so when the hazard hits in a context of social vulnerability (Oliver-Smith and Hoffman, 2002). The case of hurricane Katrina provides an example particularly showing this interaction between natural hazards and social vulnerabilities. Before Katrina hit, the economic situation of New Orleans was dire already: the oil and gas industry had relocated to Texas, and economic opportunities, apart from tourism, were scarce. Louisiana was one of the poorest states in the United States and 28% of New Orleans' population lived below the poverty line. Amongst those, 84% were African Americans. One quarter of the population did not own a car and many had never left the city.

The city itself had grown out of the historical French Quarter alongside the banks of Mississippi, in a disorganised and anarchic fashion: as the population expanded, levees were built and flood-prone neighbourhoods were developed (Colten, 2004). Many experts had warned of the risks faced by the city in the case of a hurricane: they had pointed not only to the geographical location of the city, below sea-level and nested between the Gulf of Mexico, the Lake Pontchartrain, and the Mississippi River, but also the degradation of the wetlands, the fragile and poorly maintained levees, or the alarming state of the pumps. They were not heard and sometimes not even listened to (van Heerden and Bryan, 2006).

Although the disaster brought devastation to the whole city, some neighbourhoods were more affected than others, and some segments of the population were more affected than others as well. These segments were those with the highest rate of social vulnerability which could not be reduced to the sole criterion of poverty. Laska and Hearn Morrow have shown that those most affected by the disasters were poor, black, female, old, renters and in poor medical condition (Laska and Hearn Morrow, 2006). Most of the time, these

F. Gemenne (✉)
United Nations University, Bonn, Germany
e-mail: francois.gemenne@sciences-po.org

T. Afifi, J. Jäger (eds.), *Environment, Forced Migration and Social Vulnerability*,
DOI 10.1007/978-3-642-12416-7_3, © Springer-Verlag Berlin Heidelberg 2010

factors were cumulative, amounting in a high rate of social vulnerability, far higher than the national American average.

Such inequalities in the face of the hurricane were revealed and exposed during the evacuation: most of those who did not evacuate and were stranded in the flooded city belonged to the categories mentioned above. These inequalities were not only exposed during the evacuation but also during the resettlement process and the rebuilding of the city; in this chapter, however, I restrict my analysis to the evacuation patterns and the hurricane's immediate aftermath, based on a fieldwork conducted between December 2006 and April 2007.[1] I expose the role played by social vulnerabilities in the evacuation pattern and then look at the way the evacuation and the evacuees were characterised by the media, through the lens of the 'refugee controversy'. Finally, I analyse how the controversy relates to the revealing of these vulnerabilities and inequalities and try to draw some lessons for the current debates on the characterisation and definition of environmental migrants.

2 The Evacuation: A Sociology of Those Left Behind

The evacuation of the Gulf Coast was the largest internal displacement ever experienced in the United States: an estimated 1,200,000 people fled the Greater New Orleans area in a couple of days, a displacement whose magnitude was often compared to the Dust Bowl migration (Grier, 2005). Interestingly, both events were induced, at least partly, by environmental changes.

The evacuation plans of local authorities drew much criticism from commentators and scholars like Shughart II (2006) and Brinkley (2006). Shughart recalls that

> *Despite the alarms being sounded by LSU's*[2] *storm-trackers and a personal telephone call on Saturday from the director of the National Hurricane Center, warning him of the seriousness of the threat New Orleans faced, Mayor C. Ray Nagin did not issue an order to evacuate the city, until Katrina was within 48 hr of making landfall and did not make evacuation mandatory until late Sunday morning, when fewer than 24 hr remained (Shughart, 2006: 37).*

The Governor of Louisiana, Kathleen Babineaux Blanco, was also widely accused of being too slow to react and deferring much of her power to Mayor Nagin when it came to the evacuation. In the final report of the committee set up by the House of Representatives to investigate the government failures before and after hurricane Katrina, it is noted that

> *Despite warning* 56 *hours before landfall, Governor Blanco and Mayor Nagin delayed ordering a mandatory evacuation in New Orleans until* 19 *hours before landfall.* (...) *The*

[1] Funding for this fieldwork was provided by the Belgian National Fund for Scientific Research (FNRS).

[2] Louisiana State University.

> *failure to order timely mandatory evacuations, Mayor Nagin's decision to shelter but not evacuate the remaining population, and decisions of individuals, led an incomplete evacuation* (Select Bipartisan Committee to Investigate the Preparation for and Response to Hurricane Katrina, 2006: 2).

This criticism, however, has to be nuanced. Most of the respondents I interviewed were mainly critical of the federal administration, and the Federal Emergency Management Agency (FEMA) in particular, while many of them praised local authorities for the way the evacuation was handled. Even though this observation has no statistical value, I observed that the poorest respondents had a stronger tendency to blame the federal administration and that the judgement on the administration was – unsurprisingly – strongly correlated with the experience of the evacuation: a horrendous experience was more likely to be associated with criticism towards the administration.

2.1 *The Two Stories of the Evacuation*

There are actually two stories of the evacuation (Kiefer and Montjoy, 2006: 122). The first one is the relative success of the traffic flow management for those who evacuated by themselves – the vast majority of the population. This success was achieved in large parts thanks to a contraflow system that was implemented on the I-10 motorway.[3] The contraflow allowed motorists to use both sides of the motorway, thus reducing considerably the potential traffic jams, even though several respondents mentioned that it had taken them more than 10 h to drive to Baton Rouge during the evacuation, instead of the usual 1.5 h. Boyd also notes that more than 1 million moved out of harm's way within 48 h, a number far above the 65% exit capacity of the city. This number represented 90% of the population at risk, but only 65% of those who did not own a car (carless population) (Boyd, 2006).

The other story of the evacuation is the one that drew most attention: around 15% of the population of New Orleans, representing about 10% of the population at risk, did not evacuate, because they were unable to do so and chose not to. About 110,000 people rode out the storm at home or in a friend's or relative's house, including 70,000 in New Orleans (Boyd, 2006).

People stranded in New Orleans were eventually gathered in the Superdome football stadium which had previously been identified as a refuge for the stranded population. As the capacity of the Superdome was exceeded, inhabitants moved into the Convention Centre, where similar facilities had not been installed. Some stranded residents were also directly evacuated by the federal, state, and local agencies, as well as by local volunteers. Coast Guards were

---

[3] The I-10 is the principal exit route out of New Orleans and connects the city to Baton Rouge, and further to Texas.

credited with 33,000 rescues and Louisiana Wildlife and Fisheries with 22,000 rescues (Boyd, 2006).

These two stories resulted in a strange paradox: although the evacuation of the city overall was a relative success, it was the incompleteness of the evacuation, abundantly shown in the media that triggered the public uproar on the management of the disaster. The evacuation rate was far above the most optimistic predictions of evacuation planning experts (van Heerden and Bryan, 2006), but the fact that about 70,000 individuals remained stranded in the city in appalling conditions was the focus of the most vehement criticism of the disaster management. I argue that the reason for this was that the evacuation was racially and socially stratified, exposing massive inequalities in vulnerability. These inequalities had been grossly neglected, and this negligence was perceived as racism. In the above-mentioned survey conducted for the Washington Post by the Kaiser Family Foundation and Harvard University, 68% of the respondents claimed that race and poverty had influenced the speed of the rescue efforts (Survey of Hurricane Katrina Evacuees, 2005), even though the official assessment of the authorities' failures rather pointed to a lack of preparation, coordination, and initiative, without mentioning race and poverty as determining factors (Select Bipartisan Committee to Investigate the Preparation for and Response to Hurricane Katrina, 2006).

2.2 *The Social Vulnerabilities of Those Who Did Not Evacuate*

The reasons why some people did not evacuate are multiple and intertwined. A survey conducted by the Washington Post, the Kaiser Family Foundation, and Harvard University amongst evacuees in Houston identifies the main factors that affected the migration decision of those who stayed behind (Survey of Hurricane Katrina Evacuees, 2005):

- Fifty-five percent of the respondents indicated that one of the reasons why they did not leave was that they did not own a car or had no way to leave, and 36% indicated that this was the main reason why they did not leave. It is useful to mention here that roughly one quarter of the New Orleans population did not have a car, a situation highly unusual in American cities. Furthermore, internal mobility was more limited in New Orleans than in other American cities: 77.4% of the people living in New Orleans were born in the city and had lived there most of their lives (Fussell, 2006). Some had never been out of the State of Louisiana or even the city. A group of the respondents I met did not evacuate because they just did not know where to go.
- Another prominent reason that prevented some from evacuating the city was the idea that the storm 'would not be as big as it was'. Sixty-four percent of the people surveyed mentioned this as a reason not to leave, and for 29% of them, it was the primary reason. This might sound surprising, given the numerous warnings and intense media coverage of the hurricane. However,

as stated above in the first section, one must keep in mind that New Orleans was often in the path of hurricanes, and evacuations were almost an annual routine, resulting in a 'cry wolf' effect. As expressed by one interviewee:

I didn't want to evacuate, I thought it would be like every year, a lot of hassle, and the traffic jams, all for nothing... My children finally convinced me that I had to evacuate, and they came to pick me up at 5 in the morning. I had only packed a few things, because I thought we'd be back the day after, as always.

- Some were also unable to leave (22% of those surveyed) or had to care for someone who was unable to leave (23% of the respondents). Evacuation procedures lacked provisions for those who were ill or disabled. At a conference[4] on the evacuation of the carless population, many participants pointed out the need for the city to compile a database of all its disabled residents, in order to be able to organise their evacuation in case of a new hurricane alert. A girl in her 20s that I interviewed mentioned how she had decided to ride the storm in her home at Kenner,[5] because she refused to leave behind her old neighbour who was unable to evacuate and – apparently – not being taken care of by the authorities. The situation in hospitals was very difficult: about 250 residents were stranded in the city's Memorial Hospital, amongst which 34 eventually died as a result of Katrina (Curiel, 2006: 2067). Some medical personnel chose to administer some patients with lethal doses of morphine, knowing that they had no chance of being rescued or that their medical condition would not sustain an evacuation procedure. Their trial was widely discussed and meditated, and serves as a perfect illustration of the disarray in which medical personnel were left.
- Other reasons included a too long wait (42%), a worry that the house would be burgled or possessions stolen (27%), a refusal to abandon the pets (9%), or simply a refusal to leave (37%).[6] Pets were not allowed in shelters. An interviewee explained that she – like some other residents – did not want to leave her dog behind. When she saw the water rising in her house, she decided to borrow her neighbours' canoe and canoed to the French Quarter, since she had heard in the news that this part of the city was not flooded. She then roamed into a friend's unoccupied house and stayed there for a few days. Another interviewee did not want to leave his French Quarter house unoccupied, so he bought himself a gun and rode out the storm in his house.

Such a survey, however, fails to capture other reasons that can explain why some segments of the population did not evacuate. Laska and Hearn-Morrow (2006: 10) have shown how social vulnerabilities were concentrated in New Orleans: these factors included poverty, race, lack of personal vehicle, and

---

[4] National Conference on Disaster Planning for the Carless Society, held at the University of New Orleans on 7–8 February 2007.

[5] A suburb of New Orleans, where the international airport is located.

[6] Multiple answers were allowed.

rented housing. Evacuation required money for food, gas, and lodging, and many poor families were unable to afford the expenses. Furthermore, the hurricane struck at the end of the month: many of the poorest residents were awaiting their paycheck, leaving even less resources available for their evacuation (Fussell, 2006).

John Beggs, professor of sociology at Louisiana State University, further insists on the role played by the social environment in explaining evacuation patterns.[7] He points out that the individuals who are most isolated are less likely to leave, since 'isolation is crippling their mobility'. This observation was confirmed by many interviewees, particularly amongst elderly people. Many of them, unwilling to leave, were convinced to do so by their children, grandchildren, or friends. Previous studies have shown a strong correlation between age and evacuation decision (Gladwin and Peacock, 1997), and this correlation could clearly be observed here.

In that regard, it should also be noted that those who had friends or relatives living outside New Orleans were more likely to evacuate: 'evacuees of all income levels prefer to stay with family or friends; thus, having kin or social networks outside the threatened area, but not too far, is an important factor encouraging evacuation' (Laska and Hearn Morrow, 2006: 10). Those without social networks were the last to evacuate (Fussell, 2006).

Finally, the role of information should not be downplayed. Several publications have already pointed out the importance of the access to meteorological information in order to minimise the damage sustained (Watkins, 2007). In the case of Katrina, although there was wide media coverage before the hurricane made landfall, only 73% of the respondents to the above-mentioned survey were aware of the evacuation order before the hurricane hit (Survey of Hurricane Katrina Evacuees, 2005). Amongst those who heard the evacuation order (mostly through television and/or radio), 32% asserted that the evacuation order did not give clear information about how to evacuate. In general, many factors influenced the decision and ability to evacuate, and these factors were often combined. Although most of the evacuation was rather successful and carried out before the hurricane made landfall, a fraction of the New Orleans population remained trapped in the city. Those trapped in the city were mostly black, poor, renters, and/or carless. These social vulnerabilities cumulated and resulted in environmental vulnerabilities which were directly dependent upon socio-economic factors. Laska and Hearn-Morrow (2006: 11) correctly observe that 'evacuation decisions were shaped by income, age, gender, access to information and transportation, health and physical mobility, occupations, and social networks'. Those who could not evacuate on time were neglected by authorities, not only before the hurricane hit, but also in its aftermath. The media coverage focused mostly upon this category. This is discussed in the next section.

[7] Interview with the author, Baton Rouge, March 2007.

3 How the Media Shaped the Catastrophe

In the immediate aftermath of the disaster, the media were highly praised for their role in assisting the recovery efforts. The local radio station WWL, an affiliate of CBS, took phone calls from people trapped in their house and helped rescuers locate them. The local newspaper Times-Picayune did not miss a single day of the crisis, despite its offices being flooded. The whole staff relocated in the Communication Department of Louisiana State University at Baton Rouge, and the paper continued to be edited on the Internet. CNN dispatched its reporter Anderson Cooper to New Orleans very early after the flood and played an instrumental role in alerting the general public to the drama that was unfolding in New Orleans. In the absence of any local or state officials, the media were the only link between the trapped residents and the external world. I first discuss the role played by the media in shaping the catastrophe and then discuss how the refugee controversy developed.

3.1 An Image for the Victims

Given their role in the aftermath of the disaster, it is not surprising that the media played a central role in shaping the language characterising the disaster and its victims. Many studies have shown the process and mechanisms that lead to this characterisation (Hopkins, 2007b; Sommers et al., 2006; Tierney et al., 2006). Most of the reporting was done with the people who had not evacuated New Orleans or had relocated in Houston, and less attention was given to those who had fled earlier and relocated elsewhere. Soon, the image of Katrina's victims became black and poor, as clumsily expressed by news anchor Wolf Blitzer on CNN: 'You simply get chills every time you see those poor individuals... so many of these people, almost all of them that we see, are so poor, and they're so black' (Brinkley, 2006: 204).

Through this characterisation, the media also constructed social perceptions, sometimes different from the perception of the victims themselves: the National Guards were often pictured hugging babies or rescuing people, but they were perceived as ominous by the very people they were supposed to help.

The focus rapidly shifted from the levees and the floods to the social unrest such as the looting and shooting that were supposed to have taken place in the city. The lack of food supplies in New Orleans strongly contrasted with the fully stocked shelves of stores and supermarkets that had all been locked down. Soon enough, stranded residents broke into the stores, looking for food and supplies. These actions were often described by the media as 'looting', leading to a 'criminalisation of New Orleanians in Katrina's wake' (Kaufman, 2006). Generally, it was widely acknowledged that these reports were grossly exaggerated and over-dramatised. I argue that this view also allowed for the shift that was observed in the Texan media, where evacuees in Texas were described as victims in the first days following the evacuation, then increasingly as trouble-makers and even thugs and criminals as time went on. Furthermore, Hopkins (2007a) showed that these reports had impacted negatively on the public perception towards the poor and the black, while Tierney et al. (2006) brilliantly showed the linkages between these perceptions and the handling of the crisis by the authorities. According to the authors, a common myth about disasters links the latter to social unrest, violence, and looting. They argue that 'the media's relentless adherence to disaster myths and to frames emphasising civil unrest and urban insurgency (…) had a number of immediate negative consequences'(Tierney et al., 2006: 77). The most obvious of these negative consequences was the sending of military personnel to New Orleans, with the mission of enforcing law and order, rather than bringing food and supplies. Tierney et al. (2006) argue that the conjunction of media reports and disaster myths on social unrest reinforces the political discourse for the greater involvement of the military in disaster management, as well as militarism as an ideology in the United States. As said above, the presence of the military in New Orleans, and their focus on law and order, generated much criticism.

In general, the media constructed an image for the victims of the hurricane: focusing on those who had not evacuated, they portrayed them as black and poor, and therefore stressed the social vulnerabilities and inequalities that were suddenly exposed by Katrina, although they had existed for a very long time before the storm hit. Furthermore, the media insisted heavily on highlighting the looting and social unrest and conveyed an image of the victims as vandals and trouble-makers. In doing so, they reinforced a distorted racial framework, where blackness and poverty had long been pathologised (Dyson, 2006). If the victims had no capacity to act against the images and public perceptions that were constructed, they revolted against the name they had been given: refugees.

3.2 A Name for the Victims or the Refugee Controversy

In the immediate aftermath of the hurricane, the word most commonly used to describe the victims was 'refugees' (Sommers et al., 2006: 40), but the term was unanimously rebutted by the victims themselves who insisted on being called 'evacuees' or 'survivors'. Amongst the persons I interviewed, only one agreed to being called a refugee. Interestingly, she had been a refugee before, fleeing France during the Second World War to settle in Louisiana with her husband. 'I felt I was reliving my life again', she explained 'that was exactly the same experience. I don't see what's wrong with "refugees", that's exactly what it was.' All others insisted they were not refugees, with the most common rationale for this rebuttal being that 'refugee' implied they were foreigners in their own country. 'We were transplants, not refugees', explained one evacuee, 'refugees are from foreign countries'. Refusing the 'refugee' label was also a way for many to stress the duties and obligations of their country towards them. Former New

Orleans Mayor, Marc Morial, offered the wording 'citizen refugees' without success (Gordon, 2005), and President Bush also weighed in the controversy, stating during a meeting with representatives of voluntary organisations on 6 September 2005:

> *You know, there's a debate here about refugees. Let me tell you my attitude and the attitude of people around this table: The people we're talking about are not refugees. They are Americans, and they need the help and love and compassion of our fellow citizens* White House Press Secretary Office (2005).

Shortly after the controversy erupted, many news media stopped using the word and replaced it by others, most often 'evacuees'. Mark Schleifstein of The Times-Picayune explained that he received a memo from his editor asking reporters to use the word no longer.[8] Lolis Ellie, another reporter for the same newspaper, told me that he also stopped using it, just 'because it obviously hurt people's feelings', but he felt there was no other adequate word to describe the plight of the evacuees. 'In my view, "evacuee" is not strong enough; these people had lost everything and were looking for a refuge, but I didn't want to add more suffering to their life'.[9] Other news media, on the other hand, kept on using the word, most notably CNN, The New York Times, and The Associated Press. On 7 September 2005, news anchor Lou Dobbs made the case for the continuous use of the term 'refugee' on CNN:

> *You've heard on this broadcast, by the way, several people, including Reverend Jesse Jackson and others admonish us not to use the term refugee when describing the New Orleans citizens who have had to flee their homes. Jackson and others, including President Bush, have said or implied that the term is racially insensitive. In my opinion, straight-forwardly, Reverend Jackson and President Bush are not entirely correct. The Miriam Webster Dictionary defines refugee as one who flees. (...) The president, Jackson and others apparently think that news organizations created the term refugee just to describe victims of Hurricane Katrina. Hardly. Even a cursory review of reporting of such disaster of Hurricane Andrew, the* 1993 *mid-western floods and wildfires through the west have all prompted the use of the term refugee by news organizations. I'm proud to tell you that this network has resisted others telling them how to use words. Rejecting, in fact, the United Nations suggestion that we use, instead of refugee, the expression internally displaced persons. I love that one. We'll continue here to use the term on this broadcast where we think it is most descriptive.* Dobbs (2005)

Other reasons were also put forward to dismiss the term 'refugee'; by some, it was argued that the term stripped people of their dignity (Masquelier, 2006),[10] while others, such as Reverends Jackson and Sharpton, contended that the term was racially biased. Linguist Geoffrey Nunberg (2005), in a study posted on his personal website, observed that the term had been used before to describe the people fleeing the Dust Bowl – another case labelled 'environmental migration'

---

[8] Interview with the author, New Orleans, February 2007.

[9] Ibidem.

[10] Interestingly, after the 1999 floods in Venezuela, President Chavez suggested to call the victims *dignificados*, instead of *damnificados*.

– and also that the word 'refugee' was twice more likely to be used than 'evacuee' when the word appeared within 10 words of the words 'black' or 'poor'. This led him to the conclusion that 'those disparities no doubt reflect the image of refugees as poor, bedraggled, and forlorn, and they suggest that there's a genuine basis for the impression that the word tends to single out one group, even if unwittingly'.

As Masquelier (2006: 135) points it, the unease about the word 'refugee' in New Orleans points towards a categorical void:

> *The word (...) was unsuitable to describe the plight of people who had been (...) forced to evacuate their homes prior to or in the days following the storm: they had left their homes, not their country. And yet, (...) there appeared to be no adequate substitute that would convey with enough intensity the nightmarish experience so many New Orleanians had gone through in the wake of Katrina and the dire predicament survivors (...) faced as they prepared to start life anew away from the ravaged Gulf Coast.*

Masquelier (2006) further argues that this categorical void reveals a social void in which poor New Orleanians had been confined for years: Could this claim be transposed to environmental migrants in general, often misnamed and lacking a proper terminology?

4 Conclusion: What's in a Name? Semantics and Vulnerabilities

The failure of the authorities in the aftermath of the disaster was not confined to emergency relief; they revealed a much larger social failure of the social system and exposed vulnerabilities to a country that had long been blind about them. In many ways, Katrina was a social disaster of unprecedented amplitude. In that regard, the word 'refugee' was more than just a name, but a summary of these vulnerabilities. In many ways, however, those stranded in the city experienced a refugee-like situation of stress, trauma, and despair; for many of them, the place where they had been relocated was indeed another country. Although the word 'refugee' was widely used by the media to describe all people displaced by the hurricane, the images used always focused on the poor, black residents stranded in the city, and soon the word 'refugee' equalled 'black' and 'poor', as if those who had evacuated before the hurricane belonged to another category: the 'evacuees'.

The reason why those called 'refugees' had been stranded in the city was directly dependent upon their social vulnerabilities, as I tried to show in the first section of this chapter. I argue that the reason why they rebutted the R-word was that it was perceived as a stigma for these vulnerabilities. Masquelier (2006: 741) rightly asserts that 'our failure to find a word that would describe appropriately the dire circumstances of so many Katrina victims without further victimizing them hints at a much larger failure, one that resulted over the years in the virtual disenfranchisement of a whole stratum of the U.S. population'. The lack of adequate vocabulary also reveals the lack of visibility of the vulnerabilities, the lack of a conceptual machinery to acknowledge and address them – I contend that this is a major explanation of the failures of the evacuation and relief efforts.

In the context of Katrina, two readings of the rebuttal of the word 'refugee' can be made: a refusal to not only be considered as 'second-class' citizens or foreigners in their own country, but also a refusal to admit that New Orleans residents had been failed and abandoned by their own government. This dual reading reflects a double estrangement. The first one conveys the idea that the refugee label confiscated the identity and citizenship of those it designated and confined them in a state of displacement, whereas their only wish was to return to New Orleans as soon as possible. The second reading emphasises the obligations of the state towards its citizens and conveys the anger at the lack of response by the authorities. Many interviewees insisted that their government had a duty to care about them and refused the idea of a failing state that could just abandon them and let them become refugees indeed.

Furthermore, this controversy also has its implications for the current academic and political debates on the definition and typology of environmental migration, as it reminds us how much categorisations matter, even in the most dramatic circumstances. These categorisations shape the identity of the victims, and thus one can assume that they also affect their ability to cope with the disaster. Until now, the academic community has widely considered the term 'refugee', when dealing with environmental migration, as a misnomer for legal reasons: people displaced by environmental changes are not recognised as refugees under the 1951 Geneva Convention. The controversy that erupted in the wake of Katrina invites us to reconsider the role of language and words in shaping the perception of the victims of disasters and the impact of categories on the people they are supposed to classify. In any case, it certainly shows that 'refugee' might not be a misnomer for legal reasons only.

Acknowledgements This field work was made possible thanks to a travel grant of the National Fund for Scientific Research (FNRS) in Belgium. Furthermore, I would like to acknowledge the support of Thomas Langston (Political Science Department, Tulane University), John Kiefer, and Shirley Laska (Political Science Department and CHART, University of New Orleans), as well as Marc Levitan and Ezra Boyd (Hurricane Center, Louisiana State University) for the help and facilities provided during my stay.

References

Boyd, E. (2006): Hurricane Katrina related fatalities in Louisiana. *ESRI GIS Health Conference*, Denver, CO.

Brinkley, D. (2006): *The Great Deluge: Hurricane Katrina, New Orleans, and the Mississippi Gulf Coast*. William Morrow, New York, NY.

Colten, C.E. (2004): *An Unnatural Metropolis: Wresting New Orleans from Nature*. Louisiana State University Press, Baton Rouge, LA.

Curiel, T.J. (2006): Murder or mercy? Hurricane Katrina and the need for disaster training. *New England Journal of Medicine* 355(20), 2067–2069.

Dobbs, L. (2005): *Lou Dobbs Tonight*: Cable News Network. Broadcast on 7 September.

Dyson, M.E. (2006): *Come Hell Or High Water: Hurricane Katrina and the Color of Disaster*. Basic Civitas, Cambridge, MA.

Fussell, E. (2006): *Leaving New Orleans: Social Stratification, Networks, and Hurricane Evacuation*. Social Sciences Research Council 2006. Accessed 14 May, 2007. Available from http://understandingkatrina.ssrc.org/Fussell

Gladwin, H., and W.G. Peacock. (1997): Warning and evacuation: A night for hard houses. In: Peacock, W.G., Gladwin, H., and Morrow, B.H. (ed.), *Hurricane Andrew: Ethnicity, Gender and the Sociology of Disaster*. Routledge, London.

Gordon, E. (2005): *Former New Orleans Mayor Marc Morial*. National Public Radio. Broadcast on 1 September.

Grier, P. (2005): The great Katrina migration. *Christian Science Monitor*, 12 September.

Hopkins, D.J. (2007a): Flooded communities: The Post-Katrina migration negative impact on attitudes towards the poor and African Americans. Unpublished dissertation, Department of Government, Harvard University, Cambridge, MA.

Hopkins, D.J. (2007b): Politicized change: How local reactions to the Post-Katrina migrants were shaped by the media. Unpublished paper, Center for the Study of American Politics, Yale University, New Haven, CT.

Kaufman, S. (2006): The criminalization of New Orleanians in Katrina's wake. Social Sciences Research Council 2006. Accessed 12 October, 2007. Available from http://understandingkatrina.ssrc.org/Kaufman

Kiefer, J.J., and R.S. Montjoy. (2006): Incrementalism before the storm: Network performance for the evacuation of New Orleans. *Public Administration Review* 66(s1), 122–130.

Laska, S., and B. Hearn Morrow. (2006): Social vulnerabilities and hurricane Katrina: An unnatural disaster in New Orleans. *Marine Technology Society Journal* 40(3), 7–17.

Masquelier, A. (2006): Why Katrina's victims aren't refugees: Musings on a "dirty" word. *American Anthropologist* 108(4), 735–743.

Nunberg, G. (2005): When words break down. Accessed 14 May, 2008. Available from http://people.ischool.berkeley.edu/~nunberg/looting.html

Oliver-Smith, A., and S.M. Hoffman. (2002): Introduction: Why anthropologists should study disasters. In: Oliver-Smith, A., and Hoffman, S.M. (ed.), *Catastrophe and Culture: The Anthropology of Disaster*. The School of American Research Press, Santa Fe, NM.

Select Bipartisan Committee to Investigate the Preparation for and Response to Hurricane Katrina. (2006): A failure of initiative. The Final Report of the Select Bipartisan Committee to Investigate the Preparation for and Response to Hurricane Katrina. House of Representatives, Washington, DC.

Shughart, W.F., II. (2006): Katrinanomics: The politics and economics of disaster relief. *Public Choice* 127(1–2), 31–53.

Sommers, S.R., E.P. Apfelbaum, K.N. Dukes, N. Toosi, and E.F. Wang. (2006): Race and media coverage of hurricane Katrina: Analysis, implications, and future research questions. *Analyses of Social Issues and Public Policy* 6(1), 39–55.

Survey of Hurricane Katrina Evacuees. (2005): *The Washington Post*, Kaiser Family Foundation, Harvard University, Cambridge.

Tierney, K., C. Bevc, and E. Kuligowski. (2006): Metaphors MATTER: Disaster myths, media frames, and their consequences in hurricane Katrina. *Annals of the American Academy of Political and Social Science* 604(1), 57–81.

van Heerden, I., and M. Bryan. (2006): *The Storm. What Went Wrong and Why During Hurricane Katrina – The Inside Story from One Louisiana Scientist*. Viking, New York, NY.

Watkins, K.D. (2007): Human development 2007/2008. Fighting climate change: Human solidarity in a divided world. United Nations Development Programme, New York, NY.

White House Press Secretary Office. (2005): President meets with representatives from national voluntary organizations. *Press Release*, 6 September.

Part II
Approaches for the Study of Forced Migration

The African continent has been suffering for decades from enormous problems, such as poverty, disease, malnutrition, and low levels of literacy. Partly due to the various ethnical backgrounds in the continent, wars and conflicts have occurred and worsened the living conditions of the populations. Disputes over the limited natural resources were very often the triggers for war. Environmental degradation is leading to more problems in Africa, including internal and international forced migration. This part of the book deals with internal and international forced migration in Africa and illustrates two interesting approaches to studying the causes and consequences.

The first chapter uses a statistical approach and identifies the extent to which different factors – including conflict – determine the international migration in Sub-Saharan Africa as well as the relative importance of environmental factors as determinants of migration in the region. Overall, the statistics suggest that conflict and the quest for job opportunities are the most significant determinants of international migration in Sub-Saharan Africa. The second chapter assesses the differential health impacts of people who are forced to flee from the deteriorating environments in northern regions of Ghana to urban slums of its capital. In doing so, it introduces the term 'Solastalgia' to capture the feelings of distress that an individual or community experiences through the loss of an endemic sense of place. In this case, the approach is interdisciplinary and participatory, using interviews.

Other approaches to the study of forced migration are covered in other parts of this book, including the case study approach of the EACH-FOR project in Part V and case studies on climate change in Part III.

Chapter 4
Forced Migration from Sub-Saharan Africa: The Conflict–Environment Link

Wim Naude

1 Introduction

International migration is increasing as people flee conflict and political instability and seek better opportunities elsewhere. It is expected that climate change will give a further impetus to international migration, especially from poorer countries who will find it more difficult to adapt to and mitigate the impacts of climate change on land degradation, freshwater availability, and extreme weather events. According to the Stern Report, there could be as many as 200 million 'environmentally forced migrants' (EFMs) by 2050 due to climate change (Stern, 2006).

A significant proportion of the world's poor people reside in Sub-Saharan Africa (SSA). Over the last 5 decades, economic development stagnated in most of the continent. It is also the continent with the highest number of civil conflicts in the world. Already suffering from land degradation and freshwater scarcity, climate change is expected to intensify these problems, causing more extreme weather events, such as droughts and flooding.[1] It follows that one might expect international migration from SSA countries to become more important. Indeed, the current rate of international migration from SSA is high and has seen dramatic increases in recent years. It is the region with the second largest stock of international migrants and the region with the highest growth rate in net migration – exceeding 275% between 2000 and 2005.

Despite this, there is still no full understanding of the extent to which different factors determine international migration in SSA. According to a recent review, 'there is a relative lack of studies on international migration in SSA' (Lucas, 2006: 337). More specifically, there is a lack of empirical studies

W. Naude (✉)
World Institute for Development Economics Research, United Nations University, Helsinki, Finland
e-mail: Wim@wider.unu.edu

This chapter draws extensively on Naudé (2008).

[1] Reuveny (2007) discusses 38 cases of environmental migration events in recent years. Out of these, 15 occurred in Africa, involving more than 20 million people.

T. Afifi, J. Jäger (eds.), *Environment, Forced Migration and Social Vulnerability*, DOI 10.1007/978-3-642-12416-7_4, © Springer-Verlag Berlin Heidelberg 2010

into the determinants of international migration in SSA (Konsiega, 2007; Lalonde and Topel, 1997). Although an obvious determinant of international migration is conflict – SSA is the region with the largest numbers of refugees and internally displaced persons[2] – there has not been much empirical work to quantify the impact of conflict on international migration. Lucas (2006: 365) points out in this regard that existing studies only consider proxies of the effect of conflict and have not considered 'direct indicators of the nature, duration and intensity' of conflict. A third omission in this literature is on the relative importance of environmental factors as determinants of migration, in particular on the impact of natural disasters. There is much anecdotal evidence on the extent of and future potential for 'environmentally forced migration'. However, little empirical evidence has been forthcoming on the significance and extent of such 'environmental migration'. Rightly, Bogardi (2007) has recently asked to what extent migrations are 'environmentally driven'?

This contribution addresses these shortcomings by empirically investigating the determinants of international migration in SSA. In the next section, the panel data methodology used is explained. The empirical results are discussed in Section 3. Section 4 concludes.

2 Methodology

2.1 Estimating Equation and Estimator

Naudé (2008) uses a random utility model, following Ibánez and Vélez (2008), to estimate the determinants of international migration in SSA. This model posits the following dynamic relationship between net migration and its determinants:

$$\Delta m_{it} = \gamma_t + (\alpha + 1)\Delta m_{it-1} + \Delta x'_{it}\beta + \Delta u_{it} \tag{1}$$

For $i = 1, \ldots N$ and $t = 2, \ldots T$ and where m_{it} = net migration from country i over period t; $x_{it} = a\ 1 \times K$ vector of explanatory variables. Net migration is used and not out-migration as such, given that data on the latter are not available. The explanatory variables which are set out in Table 4.1 consist of variables representing economic opportunities, demographic and environmental pressures, armed conflict, natural disasters, and others. As indicated in Table 4.1, some of these vary over t; c_i = unobserved country characteristics that are constant over the time period and influence m_{it}; and u_{it} = a random error term with the usual properties.

[2] It is estimated that 1 out of 3 refugees worldwide are in Africa and that there are around 13 million Internally Displaced Persons (IDPs) in Africa (Black, 2004).

Table 4.1 Summary of main variables and data sources

Variable	Measures	Description	Sources of data
Migration	Net migration per 1,000 population	Difference between emigration and immigration per 1,000 of a population. Provided as annual average over 5-year periods: 1960–1965, 1965–1970, 1970–1975, 1975–1980, 1980–1985, 1985–1990, 1990–1995, 1995–2000, 2000–2005	United Nations Population Division (http://esa.un.org/unpp)
Economic opportunities and costs of migration	GDP per capita	Obtained as average GDP in constant 2000 international dollars divided by total population per year over 5-year intervals, 1960–2005. Time varying	World Bank. World Development Indicators Online
	GDP growth	Average annual growth in real GDP over 5-year intervals 1960–2005. Time varying	World Bank. World Development Indicators Online
Demographic pressures and environmental degradation	Young population	Proportion of population aged 15–24. Time varying	United Nations Population Division
	Land under irrigation	Measure of the water scarcity in a country, taken as the hectares of arable land under irrigation. With less land under irrigation a country may have less water resources and may be more dependent on rainfall/susceptible to drought. Time varying	World Bank. World Development Indicators Online
Political instability and conflict	Armed conflict	The number of years during a 5-year period when there was civil war in a country, defined as at least 25 battle deaths. Time varying	UCDP PRIO Armed Conflict dataset (http://www.prio.no/CSCW/Datasets/Armed-Conflict/UCDP-PRIO)
Natural hazards	Disasters	The total number of natural disasters in a country in a 5-year period between 1974 and 2003. Time varying. Only measure the occurrence, not the severity	Centre for Research on the Epidemiology of Disasters
Others	Landlocked	A dummy variable = 1 if a country is landlocked and 0 if not. Can be interpreted as an indicator of environmental stress, of ruggedness, as well as of openness to the world economy. Time invariant	Africa Research Programme at Harvard University (http://africa.gov.harvard.edu)

The preferred estimator in this case is a dynamic panel data estimator – specifically the 'system-GMM' (Generalised Method of Moments) estimator. This estimator allows one to deal with three issues which complicate estimation with typical linear estimators, such as Ordinary Least Squares (OLS) and Two Stages Least Squares (2SLS) in the present case.[3] First, one can expect important country-specific effects (c_i) to operate, which may cause omitted variable bias. Second, the relationship between net migration and some of its expected determinants is complex, with the possibility of reverse causality. For instance, while population pressure might lead to out-migration, the very occurrence of out-migration may act as a valve, relieving population pressure in a subsequent period. Similarly, net migration may affect economic opportunities by influencing GDP growth through changes in skilled labour. Third, migration is a dynamic process. It was noted that in the case of SSA past levels of migration may influence current levels due to either persistence effects (networks or 'family and friends') or instability (returning migrants). To allow for these, lagged values of net migration are included in Eq. (1).

2.2 Variables and Data

The variables, data, and data sources are summarised in Table 4.1. The dependent variable in subsequent regressions is net migration, i.e. the difference between the total number of immigrants and emigrants. These data, made available by the UN Population Division, have been taken for 45 SSA countries,[4] and are expressed as an annual average for the 10 five-year periods ending in 1960, 1965, 1970, 1980, 1985, 1985, 1990, 1995, 2000, and 2005.

A shortcoming in this chapter that has to be acknowledged up front is the quality and quantity of data – specifically data on net migration. A first aspect already noted is the lack of gross migration data on a country level. A second aspect is of course the use of country-level data itself. The aggregated nature of migration data on a country level and the fact that countries experience both inflows and outflows of migrants in any given year, reflecting simultaneous push and pull factors, suggests that lower-level data, say on the household level, would be more appropriate. Third, as was mentioned, there are difficulties in the accurate measurement of migration, so that the data used may not reflect

[3] A discussion of these various estimators, routinely used in econometrics, falls outside the scope of this chapter. The interested reader can be referred to, for example, Wooldridge (2008) for an overview.

[4] The countries included are Angola, Benin, Botswana, Burkina Faso, Burundi, Cameroon, Cape Verde, Central African Republic, Chad, Comoros, DR Congo, Rep. Congo, Côte D'Ivoire, Equatorial Guinea, Eritrea, Ethiopia, Gabon, The Gambia, Ghana, Guinea, Guinea-Bissau, Kenya, Lesotho, Liberia, Madagascar, Malawi, Mali, Mauritania, Mauritius, Mozambique, Namibia, Niger, Nigeria, Rwanda, Sao Tomé and Principe, Senegal, Sierra Leone, Somalia, South Africa, Sudan, Swaziland, Tanzania, Togo, Uganda, Zambia, and Zimbabwe. Due to data limitations Equatorial Guinea, Mayotte, and Seychelles are excluded.

the full extent of international migration. Fourth, existing data do not capture differences between short-term, temporary, or permanent migration, destination countries, or the composition of migrants (Lucas, 2006).

Data on the various independent variables are grouped in Table 4.1 according to the major types of determinants identified in the previous section: economic opportunities, environmental degradation and demographic pressures, political instability and conflict, natural hazards, and others.

As can be seen from Table 4.1, GDP per capita and GDP growth are used as indicators of economic push and pull factors. GDP growth is generally used in the literature as a proxy for employment opportunities (e.g. Hatton and Williamson, 2001; Lucas, 2006).

Environmental degradation and demographic pressures are measured here by the degree of water scarcity and the proportion of a country's population that is young (between 15 and 24 years of age), following Hatton and Williamson (2001) in the latter. In this study, the percentage of land under irrigation is taken as a (imperfect) proxy for water scarcity. The reasoning is that in countries with less irrigation, there are perhaps less water resources available and also a greater dependence on rainfall. Other possible indicators, such as the water stress index (see Ohlsson, 1999) or the Environmental Vulnerability Index (EVI) of SOPAC (Pacific Islands Applied Geoscience Commission), were also considered, but these indicators are only available for a single year – and were not found to result in any improvement in the results.

Population density is a potentially important determinant affecting both economic opportunities and environmental stress. Population density can therefore be seen as a proxy for land scarcity (Raleigh and Urdal, 2007) and real wage pressure (higher population density can reduce wages) (Hatton and Williamson, 2001, 2002). A potential problem when using population variables in a regression model on net migration is that population variables are likely to be endogenous. For example, migration is likely to affect population variables, so that there is likely to be strong bi-directional causality between such variables. This consideration, as well as the fact that the UN Population division makes use of population growth rates in estimating net migration rates,[5] has influenced the decision in this study to not use population growth rates as an independent variable in the regression analysis.

As far as political instability and conflict determinants are concerned, Table 4.1 shows that these are measured directly by the number of years during the 5-year period characterised by armed conflict (which resulted in more than 25 battle-related deaths). It is generally accepted that armed conflict is an important determinant of out-migration, and one would expect to find a significantly negative relationship between violent conflict and net migration in the sample.

The effect of being subject to natural hazards is measured by the number of natural disasters that have taken place in a country during the period – bearing in

[5] There is a very strong positive linear relationship between population growth rates and net migration rates in the sample. The correlation coefficient between these is 0.81.

mind that not all natural hazards result inevitably in natural disasters. The data are taken from the Centre for Research on the Epidemiology of Disasters (CRED).

Finally, Table 4.1 indicates that a category of 'other' variables is also included in the regression analysis, which includes whether or not a country is landlocked. People in landlocked countries might have more difficulties in migrating internationally, if their aim is to migrate out of Africa. Also, to the extent that landlocked SSA countries may be experiencing more economic and environmental difficulties, migration from landlocked to coastal countries may be significant.

3 Regression Results

Table 4.2 contains the 'system-GMM' regression results. It reports the normal Z-values (column 2), the Z-values obtained from estimating robust standard errors (column 3), and the coefficients from the two-step system-GMM estimator (column 4). It can be seen from the diagnostics that the overall specification seems to be sound.

The results show that the variables that are consistently significant across columns 2–4 are past levels of net migration, GDP growth, and armed conflict. Population density is significant when the one-step (non-robust) estimator is used and natural disasters when the two-step estimator is used. The signs on the coefficients are as was expected.

Forced migration due to conflict and the quest for job opportunities appear to be the most significant determinants of international migration in SSA. An additional year of conflict will raise net out-migration by 1.35 per 1,000 and an additional 1% growth will reduce net out-migration by 1.31 per 1,000. The latter results show the importance of taking lags into consideration: without lagged GDP growth, the impact of an additional 1% growth on migration is only around 0.8 per 1,000 – more close to other estimates in the literature (e.g. Hatton and Williamson, 2001) where static estimation methods are used.

The importance of dynamics is further emphasised in Table 4.2, as it shows that past levels of migration have a significant influence on current levels. The fact that the sign on lagged migration is negative suggests not persistence in international migration flows but rather a situation where there is cyclical or return migration. This confirms the pattern of volatility of international migration in SSA that has been recognised in the literature (e.g. Lucas, 2006). It is also consistent with migration that is mainly forced.

Population density is only significant in column 2 and has a negative (but relatively small) impact on net migration. This is as was expected, given the discussion in Section 3. The reason for the negative coefficient on population density could reflect lower relative wages and/or population pressure on resources. Both reasons would result in a negative coefficient.

From Table 4.2 there is only a slight indication that natural disasters may have an influence on migration in SSA. It is statistically significant in the

Table 4.2 System-GMM estimates of the determinants of net migration in SSA (dependent variable net migration), 1965–2005

Variable	Coefficients		
	One-step	Robust one-step SE	Two-step
Intercept	−13.91 (−0.66)	−13.91 (−0.59)	−13.40 (−0.95)
Net migration (lag)	−0.24 (−2.43)**	−0.24 (−2.24)**	−0.21 (−2.46)**
GDP per capita	0.003 (1.13)	0.003 (1.05)	0.003 (1.17)
GDP growth	0.82 (6.82)**	0.82 (3.26)**	0.59 (3.25)**
GDP growth (lag)	0.49 (2.33**)	0.49 (2.86**)	0.30 (2.36)**
Population density (lag)	−0.13 (−1.70)*	−0.13 (−1.29)	0.006 (0.14)
Population 15–25 years (lag)	0.93 (0.84)	0.93 (0.81)	0.72 (0.98)
Land under irrigation	−0.03 (−0.06)	−0.03 (−0.08)	−0.30 (−1.53)
Natural disasters	−0.73 (−1.63)	−0.73 (−1.57)	−0.64 (−1.65)*
Incidence of armed conflict	−1.35 (−2.19)**	−1.35 (−2.39)**	−1.54 (−5.22)**
Landlocked	−4.30 (−1.08)	−4.30 (−0.82)	−3.26 (−0.73)
Diagnostics	*No. of observations* = 147 *No. of groups* = 41 *No. of instruments*= 19 *Sargan test of overidentifying restrictions*: χ^2 (8) = 10.73, p =0.22 (*accept null that overidentifying restrictions are valid*)	*No. of observations* = 147 *No. of groups* = 41 *No. of instruments* = 19 *Arellano-Bond test for second-order autocorrelation prob* > Z = 0.85 (*accept null of no autocorrelation*)	*No. of observations* = 147 *No. of groups* = 41 *No. of instruments* = 19 *Sargan test of overidentifying restrictions*: χ^2 (12) = 8.33, p =0.40 (*accept null that overidentifying restrictions are valid*)

Z-ratios in brackets.
*Indicates statistical significance at the 10% level.
** Indicates statistical significance at the 5% level.
Source: author's calculations.

two-step estimator, with a coefficient size which suggests that one additional natural disaster per year could lead to an increase in net out-migration of 0.6 per 1,000. Apart from population density, which could be seen as reflecting pressure on resources, other determinants related to environmental degradation and resource scarcity, such as irrigation and water stress/environmental vulnerability index (not reported), were not found to be significant. However, these variables may affect conflict and job opportunities (GDP growth) and as such have an indirect impact on migration.

To investigate this further, Table 4.3 contains results of a probit regression for the determinants of whether a country will experience conflict in a given year or not. The dependent variable was constructed as a dummy variable which is equal to 1, if a country experienced armed conflict in a particular year and equal to 0, if otherwise. Furthermore, Table 4.4 contains system-GMM results of the determinants of the number of years of conflict (intensity) in SSA. The independent variables in these regressions are based on the conflict in SSA literature which posits 'greed and grievance' as causes of conflict (e.g. Collier and Hoeffler, 1998; Welsch, 2008). Thus, these regressions include, in addition to natural disasters, variables such as GDP, GDP growth, ethnic fractionalisation (using the index of Alesina et al., 2003), and in the case of the probit regression, the Environmental Vulnerability Index (EVI) of a country (alternatively the proportion of land under irrigation was also used, but turned out to be insignificant as it was in the system-GMM estimation). Moreover, the rate of net migration is included to test the possibility that migration may influence conflict in a country.

Table 4.3 shows that low GDP per capita, ethnic fractionalisation, and the number of natural disasters are the most significant explanatory variables predicting whether or not conflict will break out in SSA. Countries with a higher GDP per capita have less probability of being at civil war; conversely, countries with higher degree of ethnic fractionalisation have a higher probability.

These results are consistent with the extensive literature on civil conflict in SSA. What is novel in the present results is the significant influence of natural

Table 4.3 Probit regression results for the likelihood of armed conflict in SSA

Variable	Coefficient
Net migration	−0.020 (−1.03)
Natural disasters	0.05 (1.86)*
GDP per capita	−0.0003 (−1.83)*
GDP growth	0.01 (0.35)
Ethnic fractionalisation	0.91 (1.78)*
Environmental vulnerability index	0.002 (0.82)
Pseudo R^2	0.06

Z-ratios in brackets.
*Indicates statistical significance at the 10% level.
**Indicates statistical significance at the 5% level.
Source: author's calculations.

Table 4.4 System-GMM regression results for the determinants of the intensity of armed conflict in SSA (dependent variable = number of years of civil conflict)

Variable	Coefficient (Robust SE)
Intercept	−7.7 (−1.57)
Incidence of armed conflict (lag)	0.75 (3.30)**
Natural disasters	0.02 (0.31)
GDP per capita	−0.0005 (−0.82)
GDP growth	−0.08 (−4.08)**
Ethnic fractionalisation	9.04 (1.45)
Net migration rate (lag)	−0.006 (−0.22)
Land under irrigation	0.12 (1.21)
Landlocked	4.55 (2.31)*
Diagnostics	*No. of observations* = 154 *No. of groups* = 41 *No. of instruments* = 17 *Arellano-Bond test for second-order autocorrelation prob* > *Z* = 0.85 (*accept null of no autocorrelation*)

Z-ratios in brackets.
*Indicates statistical significance at the 10% level.
**Indicates statistical significance at the 5% level.
Source: author's calculations.

disasters. This may be consistent with the hypothesis that natural disasters can act as a 'trigger' for conflict over scarce resources (Hendrix and Glaser, 2007). Calculating the elasticity from Table 4.3 indicates that an additional 1 disaster per annum (the average in the sample was 3.6) raises the probability that the country can fall into civil conflict by 1.75%.

Table 4.4 shows that as far as the intensity of civil conflict is concerned (measured by the number of years of conflict), GDP growth reduces the intensity of conflict and conflict in the previous year (lagged conflict) prolongs conflict (i.e. there is some persistence). Being landlocked also raises a country's conflict duration. This may reflect the fact that if institutions are lacking and conflict over natural resources does take place, it has a particularly detrimental impact on countries with 'low degrees of openness', such as Africa's landlocked nations (Arezki and Van der Ploeg, 2007). Table 4.4 furthermore indicates that the number of disasters does not affect the duration/intensity of conflict – supporting the notion of natural disasters, rather as a 'trigger' for conflict. Also, as in Table 4.3, the net migration rate (here lagged to avoid problems of reverse causality) has no significant effect on conflict.

Finally, Table 4.5 reports regression results on the effect of natural disasters, environmental degradation, and resource scarcity on GDP growth (which is generally seen as a proxy for job opportunities). The purpose of the results in Table 4.5 is not to provide a comprehensive analysis of the determinants of GDP growth in SSA (that would require a separate chapter) but rather to determine whether or not natural disasters, environmental degradation, and resource scarcity might have a significant impact on GDP growth. Thus, the

Table 4.5 System-GMM regression results for the determinants of GDP growth in SSA

Variable	Coefficient (robust SE)	Coefficient	Coefficient (robust SE)
Intercept	6.63 (1.50)	−111.69 (−1.33)	−111.69 (−0.19)
GDP growth (lag)	−0.45 (−3.31)**	−0.54 (−3.58)**	−0.54 (−5.24)**
GDP per capita	−0.003 (−2.67)**	−0.004 (−1.85)*	−0.004 (−3.73)**
Natural disasters	−0.212 (−0.70)	−0.09 (−0.29)	−0.09 (−0.25)
Natural disasters (lag)	−0.29 (−1.67)*	−0.11 (−0.42)	−0.11 (−0.54)
Water stress	−1.30 (−1.42)	−1.76 (−0.55)	−1.76 (−0.08)
Crop land under irrigation	0.16 (0.47)	−0.11 (−0.30)	−0.11 (−0.56)
Population density	0.01 (0.30)	−0.02 (−0.27)	−0.02 (−0.34)
Population growth	0.33 (0.55)	0.35 (0.37)	0.35 (0.43)
Arable land	0.36 (1.14)	0.83 (1.86)*	0.83 (3.11)**
Pre-colonial centralisation		159.8 (2.02)*	159.8 (0.28)
Slave exports		−0.000 (−0.1.84)*	−0.000 (−0.26)
Malaria risk		41.97 (1.03)	41.97 (0.16)
Armed conflict		−0.44 (−1.23)	−0.44 (−1.81)*
Armed conflict (lag)		0.05 (0.14)	0.05 (0.22)
Net migration (lag)		0.02 (0.24)	0.02 (0.38)
International migrant stock		−0.16 (−0.44)	−0.16 (−0.51)
	No. of observations = 100 *No. of groups* = 35 *No. of instruments* = 24 *Arellano-Bond test for second-order autocorrelation prob* > *Z* = 0.74 (*accept null of no autocorrelation*)	*No. of observations* = 84 *No. of groups* = 30 *No. of instruments* = 28 *Sargan test of overidentifying restrictions*: χ^2 (11) = 23.05; *p* = 0.02 (*reject null that overidentifying restrictions are valid*)	*No. of observations* = 84 *No. of groups* = 30 *No. of instruments* = 28 *Arellano-Bond test for second-order autocorrelation prob* > *Z* = 0.68 (*accept null of no autocorrelation*)

Z-ratios in brackets.
*Indicates statistical significance at the 10% level.
**Indicates statistical significance at the 5% level.
Source: author's calculations.

system-GMM regressions reported in Table 4.5 took GDP growth as the dependent variable and the number of natural disasters, the degree of water stress (as measured by the Water Stress Index), the proportion of cropland under irrigation, the proportion of arable land in a country, and population growth and population density as the main explanatory variables. As controls, a selection of variables was included as proxies for some of the major determinants of SSA growth that have been identified in the literature. These include an index of pre-colonial centralisation as a proxy for institutional and policy capacity, obtained from Gennaioli and Rainer (2007),[6] a measure of the number of slave exports from a country between 1400 and 1900, obtained from Nunn (2007), armed conflict measures from PRIO, and as a measure of geographical constraints, the proportion of the population subject to malaria is used (see Gallup et al., 1999).

The results in column 2 of Table 4.5 show that the effect of the environmental variables on GDP growth is generally insignificant. Only the natural disasters variable is statistically significant with a one-period lag. Once institutions, geography, and conflict are controlled (columns 2–4), the only environmental variable that becomes significant is the amount of arable land. Decreases in arable land, for instance through soil degradation, will therefore lead to a decline in GDP growth. It can be seen from column 4 that armed conflict has a statistically significant negative contemporaneous association with GDP growth, consistent with bi-directional causality between conflict and GDP growth. Finally, there is no evidence from Table 4.5 that migration has a statistically significant impact on GDP growth in SSA.

4 Concluding Remarks

International migration is of growing importance to countries in SSA; the region already has the world's second highest stock of migrants, the world's highest stock of refugees, and the highest rate of growth in international migration. Much of SSA's migration is circular in nature. Hence, it tends to be volatile and results in substantial migration between countries within the continent. This reflects the importance of seasonal effects, natural disasters, and civil conflicts as potential underlying determinants of international migration.

Using panel data from 45 SSA countries spanning the period 1965–2005, this study found that forced migration due to conflict and the quest for job opportunities are the most significant determinants of international migration in SSA. Specifically, armed conflict and GDP growth have the largest impacts on international migration. The sizes of their respective impacts were found to be comparable: an additional year of conflict will raise net migration by 1.35 per 1,000 and an additional 1% growth will reduce net migration by 1.31 per 1,000. The size of the latter effect is significantly higher than that found earlier

[6] The difficulty with selecting measures of institutional quality in regressions with economic growth is the eventual endogeneity of good institutions. To overcome this, a measure such as pre-colonial centralisation can be used. Gennaioli and Rainer (2007) argue that this measure, the proportion of the population of a country that lived under a strong hierarchical system of political centralisation before colonisation is strongly correlated with the success of subsequent post-independence governments to extend control, provides services and establishes legitimacy.

by, for instance, Hatton and Williamson (2001). Environmental factors are also important. It was found that one additional natural disaster per year could lead to an increase in net migration of 0.6 per 1,000. Apart from population density, which could be seen as reflecting pressure on resources, other determinants related to environmental degradation and resource scarcity, such as irrigation and water stress/environmental vulnerability index, were not found to be significant.

It has to be emphasised that disentangling the separate effects of these reasons for international migration is difficult. They interact in complex ways. Thus, it was established that environmental variables affect conflict and job opportunities (GDP growth), and that conflict affects GDP growth. Specifically, the number of natural disasters in a country raises its probability of being in civil war. This is consistent with the view that natural disasters can act as 'triggers' for conflict over scarce resources. Also, natural disasters were found, in the absence of control variables, to have a negative impact on GDP growth, but with a lag period. Reductions in arable land, such as those occurring due to soil degradation, were also found to depress GDP growth. The main impact of environment factors on migration from SSA countries can thus be said to occur through their impact on conflict and economic growth.

References

Alesina, A., A. Devleeschauer, W. Easterly, S. Kurlat, and R. Wacziarg. (2003): Fractionalization. *Journal of Economic Growth* 8, 155–194.

Arezki, R., and F. Van der Ploeg (2007): Can the natural resource curse be turned into a blessing? The role of trade policies and institutions. *IMF Working Paper No. WP/07/55*, International Monetary Fund, Washington, DC.

Black, R. (2004): Migration and pro-poor policy in Africa. *Working Paper C6*, Sussex Centre for Migration Research, Brighton.

Bogardi, J.J. (2007): Environmental migration: Flight or choice? *Presentation at a Panel Discussion of UNU-EHS, UNDP, IOM, UNHCR, UNU-ONY at UN Headquarters*, New York, NY, 16 May.

Collier, P., and A. Hoeffler. (1998): On the economic causes of civil war. *Oxford Economic Papers* 50(4), 563–573.

Gallup, J.L., J.D. Sachs, and A.D. Mellinger. (1999): Geography and economic development. *International Regional Science Review* 22(2), 179–232.

Gennaioli, N., and I. Rainer. (2007): The modern impact of precolonial centralization in Africa. *Journal of Economic Growth* 12(3), 185–234.

Hatton, T.J., and J.G. Williamson (2001): Demographic and economic pressure on emigration out of Africa. *IZA Discussion Paper no. 250*, Bonn.

Hatton, T.J., and J.G. Williamson. (2002): Out of Africa? Using the past to project African emigration pressure in the future. *Review of International Economics* 10(3), 556–573.

Hendrix, C.S., and S.M. Glaser. (2007): Trends and triggers: Climate, climate change and civil conflict in Sub-Saharan Africa. *Political Geography* 26(6), 695–715.

Ibánez, A.M., and C.E. Vélez. (2008): Civil conflict and forced migration: The micro determinants and welfare losses of displacement in Colombia. *World Development* 36(4), 659–676.

Konsiega, A. (2007): Household migration decisions as survival strategy: The case of Burkina Faso. *Journal of African Economies* 16(2), 198–233.

Lalonde, R.J., and R.H. Topel. (1997): Economic impact of international migration and the economic performance of migrants. In: Rosenzweig, M.R., and Stark, O. (eds.), *Handbook of Population and Family Economics, volume 1B*. North-Holland, Amsterdam. pp. 799–847.

Lucas, R.E.B. (2006): Migration and economic development in Africa: A review of evidence. *Journal of African Economies* 15(2), 337–395.

Naudé, W.A. (2008): Conflict, disasters and no jobs: Reasons for international migration from Sub-Saharan Africa. *UNU-WIDER Research Paper*, Helsinki, Finland.

Nunn, N. (2007): Historical legacies: A model linking Africa's past to its current underdevelopment. *Journal of Development Economics* 83(1), 157–175.

Ohlsson, L. (1999): *Environment, Scarcity and Conflict: A Study of Malthusian Concerns*. Department of Peace and Development Research, University of Göteborg, Göteborg.

PRIO (International Peace Research Institute Oslo). (2007). Transnational entrepreneurs in an African outpost: Chinese migrants in Cape Verde. Accessed 22 May 2007. Available from www.prio.no/page/preview/9244/41184.html

Raleigh, C., and H. Urdal. (2007): Climate change, environmental degradation and armed conflict. *Political Geography* 26(6), 674–694.

Reuveny, R. (2007): Climate change-induced migration and violent conflict. *Political Geography* 26, 656–673.

Stern, N. (2006): *The Economics of Climate Change* (*The Stern Report*). Cambridge University Press, Cambridge, MA.

Welsch, H. (2008): Resource abundance and internal armed conflict: Types of natural resources and the incidence of 'new wars'. *Ecological Economics* 67, 503–513.

Wooldridge, J.M. (2008): *Econometric Analysis of Cross Section and Panel Data*. The MIT Press, Cambridge, MA.

Chapter 5
Solastalgia: Environmentally Induced Distress and Migration Among Africa's Poor Due to Climate Change

Petra Tschakert and Raymond Tutu

1 Introduction

Primarily due to a high dependence on agro-ecosystems and their vulnerability to environmental changes, Africa is one of the most vulnerable continents to climate change and variability (IPCC, 2007). Poor rural societies that are dependent on climate-sensitive resources will be the most affected, with potential negative impacts including loss of income, displacement, and internal migration. While substantial research has been conducted on health concerns related to climate change, including water-borne and insect-borne diseases, the psychological and emotional distress and pain triggered by slow-onset, creeping environmental transformations have been largely ignored. We apply the concept of environmentally induced illness, distress, and loss of belonging ("solastalgia") to climate change and internal migration. Albrecht (2005), based on large-scale open pit mining and drought in Australia, coined the term "solastalgia," combining *solacium* (solace), *nostos* (return home), and *algos* (pain) – the sadness, depression, or desperation caused by significantly altered environments. This notion of an environmentally induced psychological condition was taken up by the Australian science media the same year (Scatssoon, 2005).

The objective of this chapter is to assess the differential health impacts and trade-offs of those who are essentially forced to migrate from deteriorating environments in two northern regions of Ghana (Upper East and Upper West) to urban slums of the capital, Accra, and those who stay behind in their increasingly "pathological" homes. Our major hypothesis is that the degree of distress an individual or a community experiences is connected to the loss of an endemic sense of place. We also explore impacts that are linked to the transformation of social networks and customs resulting from migration.

We use Ghana as a case study for this work, in part, because there are few conflicts caused by political, ethnic, or religious tensions. Consequently, migration is more likely to be related to economic and environmental motivators. By

P. Tschakert (✉)
Department of Geography, Pennsylvania State University, University Park, IL, USA
e-mail: petra@psu.edu

T. Afifi, J. Jäger (eds.), *Environment, Forced Migration and Social Vulnerability*,
DOI 10.1007/978-3-642-12416-7_5,

using an interdisciplinary approach that provides a voice to migrants and displaced people, we capture the mental, emotional, and physical health aspects of climate change impacts in the context of Ghana's rapidly increasing rural–urban migration. In this chapter, we present preliminary findings from 100 semi-structured interviews with migrants and long-term residents in the squatter settlements of Accra. We begin with a review of the available literature dealing with health impacts of climate change. We then explore the climate change–migration nexus and the impact of migration on the physical and mental well-being of those who leave and those who remain in degraded environments. Afterward, we describe the context and methods of our study. Finally, we present our analysis from the first phase of our research and discuss the direction our research will take in the future.

1.1 Climate Change and Health

While health concerns related to climate change have been receiving increasing attention in policy discussions and media reporting, the health community itself has been reluctant to explore and theorize the multiple linkages between the two fields of inquiry (Campbell-Lendrum, 2005). Emerging comprehensive assessments (e.g., Menne and Ebi, 2006) have been largely limited to vector-borne diseases, mainly associated with floods or water shortages, and exposure to airborne allergens and heat waves.

According to Campbell-Lendrum (2005), one major reason that linkages between climate change and health outcomes are under-theorized and underexplored is methodological constraints. The author argues that epidemiological methods are usually more concerned with direct, compartmentalized cause–effect relationships rather than systemic stresses. They often involve large, randomized controlled studies and repeated trials that cannot be easily and effectively applied to health implications of global environmental and climatic change. Moreover, many of the existing studies that estimate impacts and responses do so on an aggregate level. Few (2007) encourages more attention to differential vulnerabilities and coping mechanisms at the micro-scale. The challenge is to understand how and why health impacts associated with climatic changes vary between individuals and groups in a society and to better connect health outcomes to social contexts by analyzing personal and communal experiences and interpretations of health and ill-health. To address and value the "local," Airhihenbuwa (2007) proposes locally grounded modes of framing and addressing public health problems and health education in Africa in general. Rather than perpetuating what he calls the "anthropologization" of African public health issues in which Africans themselves have no voice, he advocates entering the field of African public health through an African "gate." This implies a willingness to acknowledge the manifold conditions that create and shape vulnerability.

What has been largely overlooked in the study of climate-related health problems, in addition to their social contexts, are the psychological and mental implications of risk, disruption, displacement, and loss when adaptation is unsuccessful. Very little is known about the mental and emotional distress and pain caused by slow-onset, creeping changes, such as global warming. Yet, the WHO (1946) defines health as "a state of complete physical, mental and social well-being, and not merely the absence of disease or infirmity." O'Brien et al. (2004) deplore the lack of emphasis on local contexts of well-being and quality of life in climate vulnerability studies. They assert that subtle impacts that go beyond economic performance and food security to include a sense of belonging, respect, socio-cultural heritage, and control over one's destiny have often been disregarded as trivial and inconsequential. Research among poor, small-scale farmers in Senegal has shown that poor health and a disheartening sense of worthlessness, particularly among the male youth, undermine local adaptive capacity (Tschakert, 2007). It may very well be lack of belonging associated with environmental change that pushes ever more young Africans to risk their lives, crossing the Sahara or the Atlantic Ocean in search of a better life in Europe.

Finally, studies grounded in epidemiological methodology also usually assume that affected people (and animals) remain in high-exposure areas. However, mobility and migration have long been known to be adaptive strategies to climatic stressors, especially in Africa's drylands. Nevertheless, not everybody is able to escape exposure. McLeman and Smit (2004), for instance, assert that the most vulnerable are not necessarily the ones to migrate first, as they often lack access to money, family networks, and other types of capital.

1.2 The Climate Change–Migration Nexus

While laudable efforts have been made to deal, for instance, with post-traumatic stress disorder in the wake of Hurricane Katrina in the US and the 2004 Tsunami in East Asia, no scientific investigation could be found that addresses the subtle – but potentially more devastating – health impacts of climate change and displacement that millions of people, especially in Africa, already experience. Impacts of climate change on vulnerable regions are likely to displace large numbers of people and may even lead to mass migration (McLeman and Smit, 2004). Floods and droughts have already displaced more than 5 million people in West Africa since the 1970s, and significant migration continues to be observed from the dry Sahel southward and toward the coast (Hammer, 2004). Migration to major cities adds to the proliferation of urban slums encumbered by poor sanitation, rampant malaria, cholera, HIV/AIDS, and other infectious diseases and plagued by high levels of unemployment and crime. Although significant research has been done to investigate the myriad risks that displaced populations face, including risks related to food insecurity, marginalization, social disarticulation, and increased morbidity and mortality, the conceptually rich risk assessment approach has not been employed to analyze environmental/climate change-induced displacement (Cernea, 2005). Rural–urban migration generally and the related rapid rate of urbanization in Africa is furthermore expected to significantly reduce the proportion of the adult male labor force in rural communities, thereby further diminishing agricultural production, income, and adaptive capacity (Boko et al., 2007).

It is true that the linkage between population movements and environmental/climatic changes and degradation cannot always be neatly separated from social, political, and economic forces. The term "environmental refugees," although heavily contested, is often used to depict people who are driven from their homes in masses as a result of ecological destruction. In extreme cases, as expressed by Tolba (quoted in Lazarus, 1990: 4), "the options for the poor are either to flee or to stay put and starve." El-Hinnawi (1985: 4; cited in Kliot, 2004: 72) was the first to describe "environmental refugees" as a "process in which land degradation, mostly in Third World countries, has been the main factor in the migration of subsistence farmers into the shantytowns of major cities, producing desperate populations vulnerable to disease and natural disasters." Environmentally induced population migration has been linked to environmental stress and (short-term) natural disasters, cumulative, slow-onset changes, such as desertification and global warming, and permanent changes in habitat due to development projects, industrial accidents, and conflict and warfare (Lonergan, 1998). According to Adeel (2006), increasing numbers of people will migrate as climate extremes become more frequent and severe and livelihoods deteriorate.

Usually, refugee status is defined in political terms[1] and in relation to populations that migrate across national borders. Most environmental refugees are in fact expected to be in the category of "internally displaced people," defined as persons who have been forced to flee their homes suddenly or unexpectedly in large numbers, because of armed conflict, internal strife, systematic violations of human rights or natural or human-induced disasters, and who are within the territory of their own country (Cohen, 1996). This distinction is important for two reasons. First, no reliable numbers exist for internally displaced people, as many countries do not count them. Second, people displaced within their own country are not eligible for aid and medical resources that international refugees can access. Hence, while refugee-recipient countries are obliged to provide health services to refugees from other countries, the increasing health problems of their own displaced citizens tend to be overlooked.

[1] According to the office of UN High Commission for Refugees, refugees are persons who have fled their country because of a well-founded fear of persecution for reasons of race, religion, nationality, political opinion, or membership in a particular social group and who cannot or do not want to return home.

Despite the disagreement about accounting methods and best estimates for environmentally displaced people, the estimated range of numbers of affected people is disturbing. The Red Cross and Red Crescent estimated 25 million migrants from environmental stress in 2000, a number that equals those of armed conflicts (Taipei Times, 2007). According to the UK's Tearfund (Boyd and Roach, 2006), this number could be as high as 200 million by 2050. Most of these people are anticipated to come from rural areas and developing countries. Since a large proportion of this movement is from rural to urban areas within developing countries, no legal processes or structures are available for these people to seek assistance (Kliot, 2004).

2 Study Areas: Rural Districts and Metropolitan Areas in Ghana

Ghana is an ideal area for this study, as it has an established history of political stability that is rare in this region of the world. Since there are few conflicts caused by political, ethnic, or religious tensions, any migration is more likely related to economic and environmental motivators. Ghana is also a useful study location, because it exhibits varied strata of climate regimes and poverty levels. We focus on two rural regions in the North that are particularly prone to out-migration and potential *solastalgic* illness (Upper East and Upper West). They are known for their low rainfall (750–1,050 mm), high levels of poverty, lack of access to government resources, such as health care, and great reliance on agro-ecosystems that are highly vulnerable to climate change. As a control group, we select communities in the Western Region which has higher rainfall (mean of 1,600 mm), lower levels of poverty, and less indication of environmental/climatic changes. At the same time, we work in the two largest metropolitan areas that increasingly absorb migrants from the North (Accra and Kumasi, the second largest city).

In the Upper East, a recent news article (GhanaWeb, 2007) reports of suicide through hanging and self-poisoning, mainly by kerosene, insecticide powder, and dichloro-diphenyl-trichloroethane (DDT), among tomato farmers. Tomato farming is a major dry season activity in this area, yet many farmers have incurred huge debts in the business. While suicide cases apparently occur every year, cultural stigma prevents statistical recordings. A land use/cover change analysis for 1972–2000 for the most eastern part of the region reveals a 92% decline in open forest and losses in riverine vegetation (−38%), open savanna woodland (−37%), and open cultivated savanna (−8%) (Tachie-Obeng, 2006). Similarly, the Upper West, as described by Luginaah (2007), is a "land that is all dead and where all trees are gone." The author argues that environmental degradation has caused the quality of life to decline and many young men to migrate to mining and agricultural areas further south. While apparently none of those who have left the Upper West can foresee resettling at

Table 5.1 Welfare indicators for Ghana and the two northernmost regions (Upper East and Upper West)

Selected welfare indicators 2003 (in percent of total population per category)	Ghana	Ghana rural poor	Upper East rural poor	Upper West rural poor
Adult literacy rate (male)	65.8	31.1	21.2	16.3
Adult literacy rate (female)	42.3	13.7	9.7	8.6
Access to health care facilities <30 min	57.6	27.0	26.0	23.6
Child nutritional status (stunted)	32.4	37.1	32.8	23.2
Difficulty meeting food needs	12.8	22.4	47.3	30.7
Safe sanitation	55.0	9.2	0.8	4.3
Electricity	50.6	7.2	1.6	2.0
Worse household economic situation compared to 2002	51.5	52.2	49.8	61.6

Source: Ghana Statistical Service (2003).

home, remittances and domestic food aid have become the lifeline between the migrants and those who stay behind and uphold land and households.

Poverty is endemic and acute in both Northern regions. According to the most recent *Ghana Poverty Reduction Strategy* (GPRS II, 2005), the poverty incidence rates of 88–99% in Upper East (98–99% in rural areas) and 79–96% in Upper West (96–99% in rural areas) are the highest in the country. In comparison, the poverty incidence is 8% in the Accra Metropolitan Area, 10% in the Kumasi Metropolitan Area, and between 16 and 45% in the Western region (in rural areas 0–55%). Welfare indicators from the Ghana Statistical Survey (2003) show that the Upper West and Upper East are clearly below the national mean for rural poor, ranking consistently lowest in all categories compared to other regions (see Table 5.1). At 59% nationwide, the poverty rate is highest among food-crop farmers in rural savannas, which is particularly worrisome, as they include many female poor (Ghana Poverty Reduction Strategy, 2003). These high values are directly related to systemic problems in agriculture. The GPRS II (2005) cites, among others, the following issues: low productivity due to overdependency on rainfall, overreliance on traditional agricultural commodities, lack of access to credit, especially among women, low extension coverage, lack of storage facilities, lack of awareness of climate change and its impacts, creeping desertification through bush fires, and inadequate involvement of community members, particularly women, in resource management.

It comes as no surprise that more and more people from the north try to escape the poverty, agony, and hopelessness of their communities by migrating to urban areas in the south of the country. However, this trend is not without problems. The GPRS II (2005) notes that rural–urban migration has significantly contributed to slum development in Ghana. Roughly 5 million people lived in urban slums in 2001, a number that grows at a rate of 1.8% per year. A slum like Sodom and Gomorrah, located in the heart of Accra and serving as the main drain for the city into the ocean, reportedly receives up to 400 migrants every night, most of them arriving from northern districts. Kumasi, with a population of 1.4 million and an annual population growth rate of 5.2%, is the second largest city in Ghana (KMA, 2003). It increasingly attracts migrants from the poor North, many of whom try to make a living in the informal urban sector, including *kayayoos* (girl head-porters) and ever more children in petty trade (King, 2006).

The degree to which these migrants can be labeled as "internally displaced people" remains debated. In the absence of political, ethnic, or religious motivation, it can be hypothesized that many of those who seek a better life in the cities and shantytowns of the south are poor subsistence farmers from the North. Most of them are likely to have been robbed of their livelihood and well-being in their customary home environment that is being fundamentally transformed by creeping climatic changes. Many are anticipated to have crossed the threshold of what we call "pathology of home." Such displacement is a silent one, not making its way into international headlines. Officially, no internally displaced people were recorded for Ghana in 2004 (UNHCR, 2004). Nevertheless, their numbers are likely to increase as climate models for northern Ghana project a later onset (and hence shortening) of the rainy season with a 2°C temperature increase for the next 30 years, which has potentially "extensive implications for agriculture" (Kunstmann and Jung, 2005).

3 Methods

The results presented in this chapter are the fruits of the first step in a wide-ranging research project that aims to assess and address the health implications of climatic changes, migration, and social and environmental disruption among Africa's poor, by actively incorporating those who are most affected. Throughout this study, we employ a combination of participatory action research and psychometric analysis that generates a mix of qualitative and quantitative data. In the first phase of this research, 100 residents of the slums of Accra were engaged in semi-structured interviews. The results of these interviews are presented below. From this pool of participants, 20–30 migrants from the North will be selected for in-depth interviews. The next phase of our research will occur in the North where members of the migrants' families who stayed behind will be invited to assess distress at home through discussions, participatory mapping, walks, and video. The last phase of fieldwork will use this edited video footage in community science-policy workshops.

Our choice of analytic methods reflects the dual quantitative and qualitative nature of our data. In order to determine when homes become pathological (thresholds) and to diagnose *solastalgia*, we build upon the Environmental Distress Scale (EDS), an index of the bio-psycho-social cost of ecosystem disturbance, as used in Australia's open pit mining areas in the Upper Hunter Valley (Higginbotham et al., 2007). Perception of distress and thresholds for pathological homes are measured through rating scales, Likert scales, and yes/no responses, all of which allow psychometric analysis, including *t*-tests and other statistical analyses between different groups (rural versus urban; poor and vulnerable rural areas in the North versus less poor and vulnerable in the Western region; men versus women; elderly people versus youth). The qualitative data provide historical accounts and visions for the future that are needed to further engage research participants and set concrete priorities for action. Such analysis allows us to discriminate between those exposed to and affected by environmental and climatic disturbance and those who are not. Consequently, in addition to an empirical measurement for the concepts of *solastalgia* and pathological homes, the results are expected to serve as an important starting point for reducing distress and restoring ecosystem health, including engendering vital policy support.

4 Preliminary Results from Semi-structured Interviews

In the first phase of our research, we conducted semi-structured interviews with migrants living in two urban slum settlements in Accra: Old Fadama, popularly called Sodom and Gomorrah, and Nima-Maamobi. Two types of residents participated in our interviews. The first group of participants ("Migrants") was those who had migrated from the Upper East and Upper West regions and had resided in the study sites for 10–15 years. The second group of participants ("Long-Term Residents") consisted of migrants from any region of Ghana who had resided in the study sites for more than 15 years. We interviewed 38 male and 25 female Migrants aged 20–67 and 23 male and 14 female Long-Term Residents aged 27–80.

Migrants were asked to describe their neighborhood with the aim of understanding how they perceived and felt about their place of abode and also indicate the major challenges they face living there. Both Migrant and Long-Term Resident respondents residing in Old Fadama described their neighborhoods as muddy when it rains, full of wooden structures and a place with no town planning ethos. They indicated that their neighborhoods are perceived as a den of criminals and are therefore subject to frequent police raids that they found humiliating. Other descriptions given by respondents about their neighborhood in Old Fadama not only included "a comfortable place to live" but also that "it is a nasty place," and "no good roads and amenities here." The major challenge mentioned by most Migrants and Long-Term Residents in Old Fadama was the frequent outbreak of fire, which burns houses and destroys huge sums of money and property. Some respondents shared their experience of having lost properties on more than a couple of occasions as a result of fire outbreaks – obviously, a serious source of distress to residents. Some respondents attributed the frequent fires to the huge number of wooden structure houses in the community and also the negligence of some residents who use candles as their source of light. Others attributed the problem of fires to illegal electricity connections of some residents. One respondent noted that he believes more than 90% of electricity users in his neighborhood are illegally connected to the national grid. Other problems mentioned by the respondents were frequent sicknesses among children living in the slums, poor sanitation, and flooding of rooms when it rains. Respondents from Nima-Maamobi mentioned poor sanitary conditions, poor drainage systems, noise pollution from some traders, frequent theft cases, inadequate toilet facilities, and overcrowding as main problems bedeviling their neighborhoods. A major reason for distress among all migrants was the threat of ejection from the slums by both government officials and non-government officials in Accra. Additional problems mentioned by all participants in both locations included continuous quarrels among some residents, the use of the community as a hideout for some criminals, and the proliferation of mosquitoes due to the unsanitary conditions in the neighborhood.

As stressful and challenging as the slum environment sounds based on respondents' comments, they were still able to name several benefits or advantages to living in their communities. A major merit that residents of both slum communities consistently mentioned was the opportunity to make money from trading or doing any kind of business by taking advantage of the huge and dense population. In Old Fadama, respondents additionally mentioned cheap and affordable accommodation as a major advantage of living there. Another related advantage to the low rent is the non-requirement of rent advance from "landlords." There is no other part of the city of Accra where rent advance is not required of tenants, and in fact, some landlords demand as much as 3–4 years of rent advance. This is not the case in Old Fadama, and consequently, respondents indicated that this represented significant savings. Although rent advance is required of tenants in Nima-Maamobi, respondents indicated that cost of the accommodation is lower there than in many parts of the city of Accra. Respondents from Old Fadama further mentioned that their place of abode is within walking distance to their place of work – a central market (Agbogbloshie), and therefore, there are no transportation costs. Some respondents indicated that their residences are within walking distances from the main city business center, so it was a good idea to live there, irrespective of the challenges they face. Clearly, in spite of the dangers of residing in a place that is frequently engulfed by fire, the low cost of living (mainly in the areas of accommodation, transportation costs, and food) is one of the main factors influencing the perpetual stay of migrants and long-term residents in the slums.

We asked Migrants and Long-Term Residents about the reasons why they chose to permanently or semi-permanently change place of residence and why they specifically chose to relocate to the Accra area. Some of the reasons they gave were conflicts in certain parts of the north, poor agricultural yields, joining a family member, inadequate job opportunities at the place of origin, poor income, and no one to assist them in their endeavors. Respondents from the Yendi traditional area, the Nanumba District, and other parts in the Northern Region of Ghana mentioned conflicts and wars between Dagombas and

Nanumbas, Dagombas and Konkombas, Abudus and Andanis, and Komkombas and Nanumbas to be the immediate causes that made them move to the southern parts of Ghana. Some respondents from the Upper East region mentioned conflicts between two tribal groups, the Mamprusis and Kusasis, as the straw that broke the camel's back with regard to their decision to migrate. Respondents additionally indicated that conflicts prevented farmers from getting to their farms, and as a result, there were food shortages at home and in the markets. One respondent noted "... there was food shortage, because all the farmers were on the battle field." On the question of whether there was anything that made their lives in the north intolerable, apart from the wars and conflicts, some migrants mentioned family problems (for example, maltreatment from stepmothers), inadequate quick income-generating ventures, extreme poverty, destroyed assets, and inability of parents to provide basic needs. Respondents indicated that social networks in the destination communities influenced their choice to migrate, specifically to the slums of Accra. One migrant noted that, "I had a brother in Accra, which attracted me to this city. If I knew someone else somewhere, I might have gone there." Most of the migrants also mentioned the popular opinion that Accra had a lot of business and work opportunities, and if an individual is not lazy, s/he will certainly find a job to do.

We also found environmental factors to be associated with the north–south migration in Ghana. Most respondents, who were mainly farmers before they migrated, indicated that their low earnings were further depleted by poor yields that they felt resulted from climatic variability and change. As one respondent remarked "... my family members are farmers, and in that year (when he migrated), we had poor yield due to prolonged dry season. So, we had no money for anything else except food for home consumption. Hence, I had to travel." Respondents reported that rainfall was reduced and irregular, and the onset of rainfall was increasingly difficult to predict, resulting in significant crop failure. Migrants also complained that the quality and fertility of soil had depreciated, and that they attributed it to persistent farming on the same piece of land for a long period of time. They stated that the poor quality of soil could no longer produce a yield sufficient to meet the needs of their families. Although they felt that fertilizers could have increased yields, the fertilizers were unaffordable. Farming as a sole means of livelihood was no longer sustainable. Therefore, some of the migrants whose subsistence depended solely on their farms abandoned their homes in search of livelihood opportunities in the slums of Accra.

5 Conclusions and Future Research Directions

The findings from these semi-structured interviews illustrate several important points. First, as touched on in our discussion above, it is not always simple to disentangle the causes of migration. Migrants indicated that in some cases conflict motivated them to move. However, they did not seem to indicate that they fled for direct fear of death or persecution (the narrow definition of refugee status) but rather because of the effect that conflict had on their ability to sustain their livelihoods. Strictly delineating displaced persons into particular "types," in this case, would certainly obscure the complex interactions of risks and vulnerability in the home communities that motivated migrants to leave. Second, it is clear that the livelihoods of dryland farmers in the North of Ghana are sensitive to climatic variation and change, and furthermore that migration is a major strategy for dealing with endangered livelihoods. Migrants, however, face a whole new set of risks and stresses when they migrate to urban slums, including fire, disease, and insecure housing. Finally, social networks play an important role in determining the destination of migrants. Therefore, it is reasonable to expect that, going forward, we will find that social networks also figure largely in the day-to-day functioning of the slum settlements, possibly having far-reaching health and well-being implications for migrants.

In the next phase of our research, we will select 20–30 Migrants among our initial participants for in-depth interviews, depending on interest and consent for participation in their original home communities. We will also select 15 of the original Long-Term Residents of the urban neighborhoods. These respondents will be invited to participate in in-depth interviews. The goal of these interviews is fivefold: (i) provide more details on the original set of questions; (ii) produce personal timelines chronicling major events in people's lives; (iii) participate in conceptual mapping (mental models) of migration; (iv) assess perceptions of environmental change and distress in current (urban) and home community; and (v) participate in community mapping with specific questions regarding belonging and identity. Further, we will contact family members of these participants residing in the sending communities and invite them to participate in similar interviews in their communities. In addition to providing insight into personal histories and motivations for migration, these in-depth interviews are expected to provide insight into migrants' perceptions of hazards and hazard exposure; the physical, psychological, and emotional health impacts of environmental hazards; impacts on social networks and social cohesion; and attempted solutions and their success rate.

References

Adeel, Z. (2006): *Mongabay*. Looming desertification could spawn millions of environmental refugees. Africa may be able to feed only 25% of its population by 2025. Available from http://news.mongabay.com/2006/1214-unu.html, December 14, 2006.

Airhihenbuwa, C.O. (2007): On being comfortable with being uncomfortable: Centering an Africanist vision in our gateway to global health. In: *Health Education & Behavior*. vol. 34, pp. 31–42.

Albrecht, G.A. (2005): Solastalgia: A new concept in human health and identity. In: *PAN: Philosophy Activism Nature*. vol. 3, pp. 41–55.

Boko, M., A. Nyong, C. Vogel et al. (2007): Chapter 09: Africa. *Contribution of Working Group II to the Fourth Assessment of the Intergovernmental Panel on Climate Change.* IPCC, Cambridge.

Boyd, S., and R. Roach. (2006): *Feeling the Heat.* Tearfund, UK.

Campbell-Lendrum, D. (2005): How much does the health community care about global environmental change? Invited editorial. In: *Global Environmental Change.* vol. 15, pp. 296–298.

Cernea, M. (2005): Concept and method: Applying the IRR model in Africa to resettlement and poverty. In: Ohta, I., and Gebre, Y. (eds.), *Displacement Risks in Africa: Refugees, Resettlers and Their Host Populations.* Kyoto University Press, Kyoto, Japan. pp. 195–258.

Cohen, R. (1996): Protecting the internally displaced. In: UN Committee for Refugees (ed.), *World Refugee Survey.* American Council for Nationalities Service, Washington, DC.

El-Hinnawi, E. (1985): *Environmental Refugees.* UN Environmental Programme, Kenya.

Few, R. (2007): Health and climatic hazards: Framing social research on vulnerability, response and adaptation. *Global Environmental Change* 17(2), 281–295.

Ghana Poverty Reduction Strategy. (2003): *An Agenda for Growth and Prosperity.* Ghana Publishing Corporation, Republic of Ghana, Accra.

Ghana Statistical Survey. (2003): Republic of Ghana, Accra.

GhanaWeb. (2007): *Frustrated Tomato Farmers in Upper East Attempt Suicide*? http://www.ghanaweb.com/GhanaHomePage/NewsArchive/artikel.php?ID=122961. April 23, 2007.

GPRS II. (2005): *Growth and Poverty Reduction Strategy* (2006–2009). Republic of Ghana, Accra.

Hammer, T. (2004): Desertification and migration: A political ecology of environmental migration in West Africa. In: Unruh, J.D., Krol, M.S., and Kliot, N. (eds.), *Environmental Change and its Implications for Population Migration.* Kluwer Academic Publishers, Dordrecht. pp. 231–246.

Higginbotham, N., L. Connor, G. Albrecht, S. Freeman, and K. Agho. (2007): Validation of an environmental distress scale. *EcoHealth* 3, 245–254.

IPCC. (2007): *Impacts, Adaptation and Vulnerability.* Working Group II, Summary for Policymakers, April 13, 2007, pp. 22.

King, R. (2006): Fulcrum of the urban economy: Governance and street livelihoods in Kumasi, Ghana. In: Brown, A. (ed.), *Contested Space: Street Trading, Public Space, and Livelihoods in Developing Cities.* Intermediate Technology Publications Ltd, Warwickshire.

KMA. (2003): *Development Plan for Kumasi Metropolitan Area, 2002–2004.* Government of Ghana, Kumasi.

Kliot, N. (2004): Environmentally induced population movements: Their complex sources and consequences. In: Unruh, J.D., Krol, M.S., and Kliot, N. (eds.), *Environmental Change and its Implications for Population Migration.* Kluwer Academic Publishers, Dordrecht. pp. 69–99.

Kunstmann, H., and G. Jung (2005): Impact of regional climate change on water availability in the Volta basin of West Africa. *Regional Hydrological Impacts of Climatic Variability and Change*, Proceedings, Foz de Iguaçu, Brazil, April, 2005. IAHS Publ. 295.

Lazarus, P. (1990): New strangers at the door. *Refugees.* No. 81, December, pp. 14–15.

Lonergan, S. (1998): The role of environmental degradation in population displacement. *Environmental Change and Security Project Report* 41, 5–15. Spring.

Luginaah, I. (2007): *Environment, Migration, Health and Food Security in the Upper West Region of Ghana.* Paper presented at the Annual Meeting of the Association of American Geographers, San Francisco, April 19, 2007.

Menne, B., Ebi, K.L. (eds.) (2006): *Climate Change and Adaptation Strategies for Human Health.* Steinkopf Verlag, Darmstadt.

McLeman, R., and B. Smit (2004): Climate change, migration and security. *Canadian Security and Intelligence Service Commentary*, No. 86.

O'Brien, K., S. Eriksen, A. Schjolden, and L. Nygaard (2004): What's in a word? Conflicting interpretations of vulnerability in climate change research. *CICERO Working Paper* No. 04.

Scatssoon, J. (2005): Drought prompts new type of mental stress. *Health & Medical News,* May 24, 2005.

Tachie-Obeng, E. (2006): Land use/land cover assessment, Ghana. Prepared for the final project report for SEMSOC, December 2006.

Taipei Times. (2007): Enter 'environmental refugees'. *Taipei Times*, February 6, 2007.

Tschakert, P. (2007): Views from the vulnerable: Understanding climatic and other stressors in the Sahel. *Global Environmental Change* 17(3–4), 381–396.

UNHCR. (2004): *Statistical Yearbook: Trends in Displacement, Protection and Solutions.* UN High Commission for Refugees, Geneva, Switzerland.

World Health Organization (WHO). (1946): *Constitution of the World Health Organization.* WHO, Geneva, Switzerland.

Part III
Climate Change and Migration

Climate change as a result of human activities is already being observed in many parts of the world, and international policy responses to reduce the causes and deal with the consequences are being negotiated very slowly. In addition to an increase in the global average temperature, rainfall is becoming more variable, extreme events (e.g. floods and droughts) are becoming more frequent and more intense, and sea level is rising. Several islands and coastal populations are threatened by displacement. A reduction in agricultural production due to unreliable rainfall, which could in turn lead to more hunger and poverty, could also be an important cause of migration.

This part of the book is concerned with the impact of climate change on migration. The first chapter synthesises existing empirical results and shows that global warming could lead to major forced displacements. It calls for treating the term 'climate or environmental refugee' with care, since many of the empirical studies show that due to the number of factors involved, no climatic hazards *inevitably* result in migration. Importantly, the chapter also points to the possible policy responses to this issue.

The second chapter draws a direct link between climate change and forced migration of indigenous communities in Alaska due to the melting of permafrost. It suggests the creation of new guidelines, based in human rights doctrine, that protect the rights of the forced migrants. It also calls for giving this category of displacement an accurate definition that helps design and implement institutional frameworks of humanitarian response.

The third chapter, based on a case study, focuses on the response of the people to climate change, particularly the response of farmers in Zacatecas (Mexico) to dry weather. It shows that the impact of climate change on internal migration in this region cannot be estimated at present, especially because people there have many other reasons to migrate, and that climate change will hardly have any impact on international migration.

Chapter 6
Climate and Migration: A Synthesis

Etienne Piguet

1 Introduction

The concern of the international community about the consequences for migration of climate change has been growing since the publication in 1990 of the first UN intergovernmental report on climate change stating 'The gravest effects of climate change may be those on human migration as millions will be displaced' (Intergovernmental Panel on Climate Change (IPCC-1), 1990: 20). In 1993, the projection by Norman Myers of 150 million environmental refugees by the end of the twenty-first century and his statement that 'the issue of environmental refugees (...) promises to rank as one of the foremost human crises of our times' further fuelled the fear of mass migrations (Myers, 1993; Myers, 1997: 175; Myers, 2002). Ten years later, Sir Nicholas Stern wrote in his review delivered to the British government on the economic consequences of global warming that: 'Greater resource scarcity, desertification, risks of droughts and floods, and rising sea levels could drive many millions of people to migrate' (Stern, 2006: 111). These gloomy predictions had a tremendous effect in the media: Filmmaker Roland Emmerich dramatised the fear of climate migration in 2004, in a scene from the film 'The Day after Tomorrow' where American citizens flee en masse from a terrible climatic disturbance from the north, only to find themselves ironically running up against the fences of the American–Mexican frontier. In 2005, Papua New Guinea appeared on the front page of the British newspaper *The Guardian* in a leading article entitled 'The First Refugees of Global Warming', announcing the decision to progressively evacuate all one thousand inhabitants of the small atoll of Carteret (Kilinailau) that was being slowly submerged by rising seas.

Notwithstanding the present media focus, the amount of systematic research on migration and climate change remains quite limited. There is

E. Piguet (✉)
Institute of Geography, University of Neuchâtel, Neuchâtel, Switzerland
e-mail: Etienne.piguet@unine.ch

A preliminary version of this chapter has been published online as a United Nations High Commissioner for Refugees Research Paper (no. 153)

T. Afifi, J. Jäger (eds.), *Environment, Forced Migration and Social Vulnerability*, DOI 10.1007/978-3-642-12416-7_6, © Springer-Verlag Berlin Heidelberg 2010

much vagueness surrounding the concepts employed, the underlying mechanisms involved, the number of persons affected and the geographical zones concerned. The use of the term ‘refugee’ by numerous authors has also led to certain confusion. It evokes the juridical status recognised by the UN Convention of 1951 referring to any person leaving his country of residence while having a ‘well-founded fear of being persecuted for reasons of race, religion, nationality, membership of a particular social group or political opinion’, although it is clear that environmental motives are absent from this list and that most displacements happen internally, within internationally recognised borders.

This chapter first tries to understand why the environmental aspect of the study of migration and refugees has been neglected up until now. It then concentrates on climate aspects by providing a synthesis of existing empirical results followed by projections of future developments.

2 A Historically Documented but Neglected Topic of Study

The links between climate and human migrations are of course not new, and population geography has, for many years, acknowledged the role played by environmental factors in explaining the history of civilisations and the emergence of cities. Thus, for mankind, the passage across the Bering Straits from America 13,000 years ago was possible due to the low sea levels of the Ice Age. According to certain authors cited by Brown (2008: 21), the Barbarian invasions of Europe around the fourth century had climatic causes as well as the Muslim expansion in southern Europe. In the same vein, the Medieval Climate Optimum which lasted between the eighth and thirteenth centuries AD allowed the colonisation of Greenland by the Vikings (Bell and Walker, 2005), and inversely, the Little Ice Age which followed forced them to leave the area. According to Le Roy Ladurie, it also led to famine migration in Northern Germany where groups of beggars fleeing bad crops were described in Magdeburg, Mayence and Lübeck in 1316/1317 (Le Roy Ladurie, 2004). Paradoxically, it also appears that drought had a positive impact on the rise of civilisations. Thus, the desertification of the Sahara and the Arabian peninsula has played an important part in the densification of the population on the banks of the Nile and consequently contributed to the birth of ancient Egyptian civilisation (Hammer, 2004: 238).

In modern times, the summer heat of 1845 and 1846 in Ireland stimulated *fungus infestans* and destroyed the potato crops, leading to the famous famine emigration towards England, the USA, Canada and Australia (Le Roy Ladurie, 2004). Similarly, the droughts of the 1930s in the plains of the American Dust Bowl forced hundreds of thousands of migrants towards California (McLeman et al., 2008), and those that struck the Sahel between 1969 and 1974 displaced millions of farmers and nomads towards the cities (Hammer, 2004). More generally, the strong link between precipitation and population density has been shown in the American Great Plains (Robinson et al., 1961), and Barrios et al. (2006) recently stated in a study of 78 countries over a 30-year period that climatic change, as proxied by rainfall, has acted significantly to change urbanisation in sub-Saharan Africa (although not elsewhere in the developing world).

With industrialisation and tertiarisation, however, the importance of the role given by population geographers and other researchers to the environment declined progressively. Already at the end of the nineteenth century, the famous 'migration laws' of E.G. Ravenstein held that economic factors were of prime importance. Their pre-eminence was almost exclusive in the theorisation of migration flows during the second half of the twentieth century (Massey et al., 1993). While certain environmental characteristics of areas studied were taken into consideration, generally, only the positive factors received any serious attention. In this respect Greenwood (1969) highlights the favourable effect of high average temperatures on internal migration in the US mainland, while Graves (1980) measures the effect of climate mildness in general on migration. Up until recently, the environment, especially when considered as a negative factor inducing forced displacements, has been absent from the study of migration on account of the dominance of what we can call an 'Economic Paradigm'. One can add that migrations linked to the environment are frequently internal and affect Southern countries. It is noteworthy that these two aspects of migration have been neglected by researchers to the advantage of studies of international migrations to Northern countries. A similar result stems from the 'Political Paradigm' that characterises the specific study of refugees (Refugee Studies). The latter give only limited attention to the link between environment and migration and often focus on political refugees as defined under the 1951 UN Convention. Consequently, it is not surprising that surveys of refugee studies only give very limited attention to environmental aspects, except for the degradation that refugees might cause themselves (Black and Robinson, 1993; Richmond, 1988; Zolberg et al., 1986).

As a result of these two paradigms among social scientists, one can note with Massey et al. (2007: 3) that in none of the disciplines of social sciences 'environmental conditions figure as salient determinants of migration decision-making' while at the same time 'many environmentalists take as an article of faith that population growth, environmental deterioration, and out-migration are fundamentally interrelated'. What is true for the impact of the environment in general holds true for the impact of climate and led to important controversy among scholars during recent years.

3 A Problematic and Controversial Concept

The term 'Environmental refugees' was popularised in 1985 as a report title for the United Nations Environment Program (El-Hinnawi, 1985).[1] It has since been widely diffused in not only both political and academic circles but has also been severely criticised (Black, 2001; Castles, 2002). Rightly highlighting the shaky empirical character and sloppy nature of most work on the subject, the critics have brought to the fore problems arising from a unidirectional link between environmental changes and migrations. For Castles 'The term environmental refugee is simplistic, one-sided and misleading. It implies a monocausality which very rarely exists in practice (...) [Environmental and natural factors] are part of a complex pattern of multiple causality, in which [they] are closely linked to economic, social and political ones.' (Castles, 2002: 5). Numerous works confirm this: when environmental deteriorations cause displacements, they are often the by-product of economic, demographic or political factors (Hugo, 1996). Moreover, vicious circle phenomena are frequent and it is not easy to isolate primary causes. Hence, population displacements will induce environmental problems that might have an effect on conflicts which themselves could exacerbate environmental deterioration, etc. (Hagmann, 2005; Reuveny, 2007). There is agreement today that natural factors are not the sole cause of migration and that the economic, social and political situation of the zone under threat can, depending on the case, increase or decrease the flow of migrants.

Another serious criticism has been addressed to the advocates of the environmental refugee concept and, in particular, to Norman Myers and his estimation of a potential 150 million refugees. They are accused of brandishing the spectre of a flood of migrants towards rich countries, thus reinforcing the position of governments having policies of closed borders who are hostile to refugees. For MacGregor (1993: 162): 'In so far as the term environmental refugee conflates the idea of disaster victim and refugee, its use brings with it the danger that the key features of refugee protection could be undermined and the lowest common denominator adopted. Because environmental can imply a sphere outside politics, use of the term environmental refugee may encourage receiving states to treat the term in the same way as economic migrants to reduce their responsibility to protect and assist.' Apart from the scientific error of oversimplifying the processes taking place, the danger here is also one of 'evacuating political responsibility by overplaying the hand of nature' (Cambrézy, 2001: 48).

Even if they might have dampened the enthusiasm of certain researchers, reservations regarding the concept of environmental refugees seem to be fully justified. They have obliged the scientific community to be mindful of the consequences of their choices of terminology and point to the need for clear definitions of the different aspects of the phenomenon. A considerable number

[1] See Kniveton et al. (2008: 29) for the complete genealogy of the term.

of terminological variants have indeed been used by researchers to refer to persons fleeing climate hazards, and more generally, environmental disturbances. While the term *environmental refugees* – sometimes *ecological refugees* or even *envirogees* (Thill, 2008) – is frequent in the English language (El-Hinnawi, 1985; Jacobson, 1988; Myers, 1993; Myers, 1997; Westing, 1992) during the 1990s, as well as in German *Umweltflüchtling* (Bächler, 1994; Richter, 1998) and in French *Réfugiés de l'environnement* (Gonin and Lassailly-Jacob, 2002), one observes the emergence of a more general terminology such as *environmental* or *ecological migrants, ecomigrants* or *ecomigrations* (Wood, 2001) or the acronyms EIPM (Environmentally Induced Population Movements) and EDP (Environmentally displaced person). In order to differentiate more precisely the nature of environmental degradation on the one hand, and the type of assistance needed on the other, Renaud et al. (2007) proposed a typology of environmental migrant crossing vulnerability and type of degradation.

Targeting more precisely *climate change migrants*, the International Organization for Migration proposes to define them as: 'Persons or groups of persons who, for compelling reasons of sudden or progressive change in the environment as a result of climate change that adversely affect their lives or living conditions, are obliged to leave their habitual homes, or choose to do so, either temporarily or permanently, and who move either within their country or abroad' (Kniveton et al., 2008: 31). This definition is well suited to designate a category of migration in an unambiguous way vis-à-vis the 1951 Convention, but – as we shall see in the following section – it might give the climate a nearly deterministic status of deus ex machina which is not confirmed in empirical research on migration.

4 Global Warming, Natural Disasters and the Progressive Evolution of the Environment

It is extremely difficult to elaborate any scientific predictions concerning the number of *environmental or climate migrants*. This is due to the multicausality of the process of migration (Boano et al., 2007) and due to the impossibility of combining climate and migration mathematical models (Perch-Nielsen, 2004; Perch-Nielsen, Bättig and Imboden, 2008). A historically informed prospective method seems more promising: the expected consequences of climate change are enumerated first, so as to establish an estimation of the populations most at risk. As pointed out by Castles (2002) and contrary to a common implicit hypotheses of numerous Cassandras of climate change migration, being at risk does not necessarily mean having to migrate. On this basis, the past experiences of similar natural events have to be documented in order to assess the proportion of the victims who might migrate temporarily or permanently over a short or long distance.

Three consequences of climate warming, projected for the end of the twenty-first century in the latest report of the IPCC, appear to be the most threatening potential causes of migration (IPCC, 2007):

- The increase in strength of tropical hurricanes and in the frequency of heavy rains and flooding due to the augmentation of evaporation correlative to temperature increase;
- The growth in the number of droughts, with evaporation contributing to a decrease in soil humidity, often associated with food shortages;
- The increase in sea levels resulting from both water expansion and melting ice.

While the first two consequences are manifested in sudden natural disasters, the third is a long-term process, which, as we will see, has very different possible implications in terms of migration. We leave aside other effects of global warming on health, conflicts or the viability of certain economic activities that may have additional consequences for migration but which remain subject to speculation.

5 Past Experiences of Natural Hazards and Migration

5.1 Hurricanes, Torrential Rains and Floods

The consequences of hurricanes and floods on population displacement are among the easiest to identify in that they manifest themselves in a brutal and direct manner. While we know approximately the number of persons affected by flooding worldwide (106 million, on average, between 2000 and 2005; according to the International Disaster Database), and by hurricanes (38 million; according to the same source), the total number of people threatened by a potential increase of this kind of disaster is, however, very difficult to estimate. No climate model is able to predict with accuracy whether or not the affected zones will be densely populated and whether the damage wrought will have tragic consequences. Apart from this methodological difficulty of forecasting, the studies carried out after such events tend to relativise their effects in terms of migration in general and long-term/long-distance migration in particular. Living mainly in poor countries, the victims have little mobility (Lonergan, 1998) and the majority of the displaced return as soon as possible to reconstruct their homes in the disaster zone (Kliot, 2004). The results from numerous research projects conducted worldwide tend to confirm this point with remarkable regularity. Thus, a synthesis of results on migration choices of victims of natural disasters displaced in 18 sites confirms – with rare exceptions – the strong propensity to return (Burton et al., 1993). In a much more indirect and incomplete fashion, studies on persons seeking asylum in Europe indicate no correlation between asylum applications and natural disasters recorded in the zones of departure. On the contrary, a significant link is confirmed regarding the political situation in these same zones (Neumayer, 2005).

On a global level, the general conclusion is thus that the potential of hurricanes and torrential rains to provoke long-term and long-distance migrations remains limited. As pointed out by Kniveton et al. (2008: 35), it is only if the affected society is highly dependent on the environment for livelihood and if human action exacerbates the environmental aspect of the disaster – as was the case with Hurricane Katrina (Reuveny, 2008) – that significant migration takes place. The level of vulnerability can be tremendously different from one region to another.

5.2 *Drought and Desertification*

In the recent past, the number of persons affected by drought has been comparable to that of victims of hurricanes and floods (146 million people, on average, between 2000 and 2005 according to the International Disaster Database). The latest report of the IPCC projects increased water shortages in Africa (74–250 million people affected in 2020) and Asia: 'Freshwater availability in Central, South, East and Southeast Asia, particularly in large river basins, is projected to decrease due to climate change which, along with population growth and increasing demand arising from higher standards of living, could adversely affect more than a billion people by the 2050s.' (IPCC, 2007: 10). Case studies, however, bring to light a contrasted picture of the consequences for migration of these kinds of change. The effect of a lack of drinking and irrigation water on migration is actually less sudden than that of the meteorological events mentioned in the previous section, and it only generates progressive departures.

On one hand, there are many well-known cases of mass population departures, in particular not only in Africa (Sahel, Ethiopia) but also in South America (Argentina, Brazil), in the Middle East (Syria, Iran), in Central Asia and in Southern Asia. Hammer (2004: 232) presents an impressive table of forced migration due to droughts and floods during the period 1973–1999 in the Sahel with a maximum of 1 million displaced persons during the drought in Niger in 1985. He affirms, 'It seems very likely that hundreds of thousands of people from rural Sahel regions are displaced every year as a consequence of environmental change and desertification' (Hammer, 2004: 234). Likewise, for Leighton (2006: 47), 'The periodic drought and desertification plaguing northeast Brazil contributed to factors causing 3.4 million people to emigrate between 1960 and 1980.'

On the other hand, many researchers strongly relativise the possible direct link existing between drought and emigration by highlighting the fact that the latter, in general, is the last resort when all other survival strategies have been exhausted. Consequently, during the 1994 drought in Bangladesh, only 0.4% of households had to resort to emigration (Smith, 2001). Other researchers hold similar views to those of the Nobel Prize winner for Economics, Amartya Sen, in remarking that famines are, in general, only marginally the direct result of environmental factors, but much more of political ones (Sen, 1981) and add that this also holds for migration. A multivariate analysis on interprovincial migrations in Burkina Faso thus shows that environmental variables, in general, only explain 5% of migrations and drought itself only 0.8% (Henry et al., 2003). In certain contexts, the effect can even be the inverse. This was the case in Mali during the drought of the mid-1980s: a reduction in international emigration was observed due to the lack of available means to finance the journey (Findley, 1994). For Kniveton et al. (2008: 34), 'Drought seems to cause an increase in the number of people who engage in short-term rural to rural types of migration. On the other hand, it does not affect, or even decrease international, long-distance moves. Second, the conceptualization of drought-affected as helpless victims who are left with no choice but to flee seems to be false.'

The general conclusion to be drawn here is that projections of increased migrations linked to drought-related phenomena remain hazardous. Consequently, it would be difficult to put a figure on the magnitude of populations at risk and the potential migrations arising from global warming-induced droughts.

5.3 Rising Sea Levels

While the first two climatic hazards mentioned do not foreshadow massive population displacements due to climate change, the potential for migration when linked to an increase in sea level is considerable. In contrast to hurricanes, floods and droughts, this phenomenon is virtually irreversible and manifests itself over a long period of time. This could make migration the only possible option for the population affected as exemplified by the historical analogue of Holland Island (Arenstam Gibbons and Nicholls, 2006). The localisation of the consequences of rising sea levels is a relatively easy task, because the configuration of coastlines, their altitude and population is well known and thus easy to integrate into Geographical Information Systems (GIS) that permit simulations and projections. Hence, it is possible to calculate – on a global scale – the number of persons living in low elevation coastal zones and threatened by rising water levels, higher tides or further-reaching waves. MacGranahan et al. (2007) define 'Low elevation coastal zones' as being situated at an altitude of less than 10 m. Even though these zones only account for 2.2% of dry land, they presently are home to 10.5% of the world population, some 602 million people of which 438 million live in Asia and 246 million in the poorest countries of the world. Other authors furnish slightly lower figures totalling 397 million people, but these, nevertheless, remain impressive (Anthoff et al., 2006).

It would be certainly an exaggeration, however, to consider that these hundreds of millions of people are all potential migrants in the near future. The latest report of the IPCC evokes, of course, the possible melting of the Greenland ice cover and the consequent 7-m rise in sea level, but this would occur over several thousand years. Of more concern to us here is the scenario of thermal expansion of the oceans. According to a future CO_2 emission estimate based on continuing economic growth but with a moderation of fossil fuel use (scenario A1B of the IPCC), there could be an increase of 0.3–0.8 m of the sea level by 2300 (IPCC, 2007). On this basis, it seems reasonable to consider populations living at an altitude of less than 1 m as being directly vulnerable during the next century. A study commissioned within the framework of the Stern report gives a considerable figure of 146 million people for this group (Anthoff et al., 2006). Mainly situated in the major river deltas and estuaries, the flood zones are particularly populated in South Asia (Indus, Ganges-Brahmaputra, etc.) and East Asia (Mekong, Yangtze, Pearl River, etc.). These two regions account for 75% of the population at risk. Certain Pacific states, such as Tuvalu or Kiribati, are – in the short-term – among the most threatened, as they are situated only centimetres above sea level. Although far less populated, they have several thousand inhabitants.

The increase in sea levels appears to be the aspect of global warming that represents the greatest direct threat for numerous populations. Contrary to hurricanes and droughts, the localisation of potential victims is possible. If no mitigation measures are taken and if no effort is made to protect the groups at risk, they will have no alternative but to emigrate.

6 Conclusion

Existing research shows that due to the number of factors involved, no climatic or environmental hazards inevitably result in migrations. Many authors note that even if disasters become more frequent in the future, political efforts and measures of protection will be able to lessen the need to emigrate, provided that the necessary financial means are made available. Even rising sea levels could be partially counteracted by the erection of dykes or the filling in of threatened zones. The Stern report is clear in this respect and states that 'The exact number who will actually be displaced or forced to migrate will depend on the level of investment, planning and resources' (Stern, 2006: 112). It nevertheless estimates these costs of mitigation to be many billion dollars.

This overview also shows that the very concept of *climate* or *environmental refugee*, because of its connotations of urgency and unavoidability, is to be handled with care. It actually evokes fantasies of uncontrollable waves of migration that run the risk of stoking xenophobic reactions or serving as justification for generalised policies of restriction for people seeking asylum. As recently stated by Massey et al. (2007: 22): '[one] should evince considerable caution in viewing "environmental refugees" as a major component of migratory streams around the world. For the most part, environmental deterioration appears to promote local searches for organic inputs or alternative employment opportunities, not a desperate search for relief in distant lands.'

At the same time, this summary clearly shows that environmental degradation can generate substantial migration flows and that global warming, in particular, could lead to major forced displacements. The latter would result principally from rising sea levels but would only progressively manifest themselves over the coming centuries with the exception of the flooding of certain islands. The increase in droughts and meteorological disasters projected by climatic models will also have impacts in terms of migration, but these will remain regional and short-termed, and are at present difficult to estimate.

The question of the international system of protection needed in order to face these challenges is thus to be answered and is all the more important because of the clear responsibility of rich countries for global warming. Simply including environmental motives in the 1951 definition of refugees seems politically unfeasible due to the very likely opposition of receiving countries. It would probably not achieve its objective of protection, as the majority of displacements takes place inside the countries affected. It would also risk threatening the coherence of an international framework of refugee protection that already has difficulty in obliging states to respect their commitments. The United Nations High Commissioner for Refugees (UNHCR), being very aware of this risk of confusion between political and non-political refugees, has always treated the idea of broadening the scope of the Convention to include environmental migrants with the utmost prudence, even if he also deems this category of the population as a possible part of his protective mandate towards Internally Displaced Persons (IDP) within states (UNHCR, 2008). As stated in 2005 by the then Under Secretary General of the UN, Hans van Ginkel, 'This is a highly complex issue, with global organizations already overwhelmed by the demands of the conventionally-recognized refugees as originally defined in 1951. We should prepare now, however, to define, accept and accommodate this new breed of refugee within an international framework' (United Nations University, 2005).

It seems that two possibilities can be envisaged with regard to this: On one hand, an increased international cooperation with a view to collective burden sharing of assistance and prevention in countries confronted with disasters, and on the other, the opening of emigration channels with the recognition of environmental push factors in subsidiary international instruments of protection, such as temporary protection schemes. This second option seems more viable for urgent cases but brings with it numerous problems, in particular the question of responsibility for the displacement of the person from the disaster zone to the receiving zone.

The discussion of these possible solutions is largely beyond the scope of this essay, but it is evident that without firm preventative action, global warming could have serious consequences in terms of forced migrations. This must be more widely recognised and stimulate scientific and political awareness.

References

Anthoff, D., P. Nicholls, R.S. Tol, and A. Vafeidis (2006): Global and regional exposure to large rises in sea-level: A sensitivity analysis. *Tyndall Centre for Climate Change Research – Working Paper* 96.

Arenstam Gibbons, S.J., and R.J. Nicholls. (2006): Island abandonment and sea-level rise: An historical analog from the Chesapeake Bay, USA. *Global Environmental Change* 16(1), 40–47.

Bächler, G. (1994): *Umweltflüchtlinge: das Konfliktpotential von morgen? = Environmental refugees: A potential of future conflicts?* Agenda Verl, Münster.

Barrios, S., L. Bertinelli, and E. Strobl. (2006): Climatic change and rural–urban migration: The case of sub-Saharan Africa. *Journal of Urban Economics* 60(3), 357–371.

Bell, M., and M.J.C. Walker. (2005): *Late Quaternary Environmental Change: Physical and Human Perspectives*. Prentice Hall.

Black, R. (2001): Environmental refugees: Myth or reality? *Perspectives New Issues in Refugee Research – UNHCR Research Paper No.* 34.

Black, R., and V. Robinson. (1993): *Geography and Refugees*. Belhaven, London.

Boano, C., R. Zetter, and T. Morris. (2007): *Environmentally Displaced People: Understanding the Linkages between Environmental Change, Livelihood and Forces Migration*. Refugee Studies Center, Oxford.

Brown, O. (2008): Migration and climate change. *International Organization for Migration – Migration Research Series 31*, Geneva.

Burton, I., R.W. Kates, and G.F. White. (1993): *The Environment as Hazard*. Guilford Press, New York, NY.

Cambrézy, L. (2001): *Réfugiés et exilés – crise des sociétés – crise des territoires*. Editions des Archives Contemporaines, Paris.

Castles, S. (2002): Environmental change and forced migration: Making sense of the debate. *New Issues in Refugee Research – UNHCR Research Paper No.* 70.

El-Hinnawi, E. (1985): *Environmental Refugees*. United Nations Environmental Program, Nairobi.

Findley, S.E. (1994): Does drought increase migration? A study of migration from rural Mali during the 1983–1985 drought. *International Migration Review* 28(3), 539–553.

Gonin, P., and V. Lassailly-Jacob. (2002): Les réfugiés de l'environnement. Une nouvelle catégorie de migrants forcés? *Revue Européenne des Migrations Internationales* 18(2), 139–160.

Graves, P.E. (1980): Migration and climate. *Journal of Regional Science* 20(2), 227–237.

Greenwood, M.J. (1969): An analysis of the determinants of geographic labor mobility in the United States. *Review of Economics and Statistics* (51), 189–194.

Hagmann, T. (2005): Confronting the concept of environmentally induced conflict. *Peace, Conflict and Development* 6(6), 1–22.

Hammer, T. (2004): Desertification and migration: A political ecology of environmental migration in West Africa. In: Unruh, J.D., Krol, M.S., and Kliot, N. (ed.), *Environmental Change and Its Implications for Population Migration*. Kluwer, Dordrecht.

Henry, S., P. Boyle, and E.F. Lambin. (2003): Modelling inter-provincial migration in Burkina Faso: The role of socio-demographic and environmental factors. *Applied Geography* 23(2–3), 115–136.

Hugo, G. (1996): Environmental concerns and international migration. *International Migration Review* 30(1), 105–131.

Intergovernmental Panel on Climate Change. (2007): *Climate Change 2007: The Physical Science Basis – Summary for Policymakers*. IPCC – Secretariat, Geneva.

Intergovernmental Panel on Climate Change (IPCC-1). (1990): Policymakers' summary of the potential impacts of climate change. *Report from Working Group II to IPCC*.

International Disaster Database. EM-DAT – OFDA/CRED – Université Catholique de Louvain, Brussels, Belgium. Available from http://www.emdat.be. 1 October, 2008.

Jacobson, J. (1988): Environmental refugees: A yardstick for habitability. *Worldwatch Paper – Washington DC No.* 86.

Kliot, N. (2004): Environmentally induced population movements: Their complex sources and consequences – A critical review. In: Unruh, J.D., Krol, M.S., and Kliot, N. (ed.), *Environmental Change and Its Implications for Population Migration*. Kluwer, Dordrecht.

Kniveton, D., K. Schmidt-Verkerk, C. Smith, and R. Black (2008): Climate change and migration: Improving methodologies to estimate flows. *International Organization for Migration – Migration Research Series 33*, Geneva.

Le Roy Ladurie, E. (2004): *Histoire humaine et comparée du climat*. Fayard, Paris.

Leighton, M. (2006): Desertification and migration. In: Johnson, P.M., Mayrand, K., and Paquin, M. (ed.), *Governing Global Desertification*. Ashgate, London. pp. 43–58.
Lonergan, S. (1998): The role of environmental degradation in population displacement. *Environmental Change and Security Project Report* (4), 5–15.
Mac Gregor, J. (1993): Refugees and the environment. In: Black, R., and Robinson, V. (ed.), *Geography and Refugees. Patterns and Processes of Change*. Belhaven, London. pp. 157–170.
MacGranahan, G., D. Balk, and B. Anderson. (2007): The rising tide: Assessing the risks of climate change and human settlements in low elevation coastal zones. *Environment and Urbanization* 19(17), 17–37.
Massey, D.S. et al. (1993): Theories of international migration: A review and appraisal. *Population and Development Review* 19(3), 431–466.
Massey, D.S., W.G. Axinn, and D.J. Ghimire (2007): Environmental change and out-migration: Evidence from Nepal. *Population Studies Center Research Report No. 07-615.*
McLeman, R., D. Mayo, E. Strebeck, and B. Smit. (2008): Drought adaptation in rural eastern Oklahoma in the 1930s: Lessons for climate change adaptation research. *Mitigation and Adaptation Strategies for Global Change* 13(4), 379–400.
Myers, N. (1993): Environmental refugees in a globally warmed world. *Bioscience* (43), 752–761.
Myers, N. (1997): Environmental refugees. *Population and Environment* 19(2), 167–182.
Myers, N. (2002): Environmental refugees: A growing phenomenon of the 21st century. *Philosophical Transactions: Biological Sciences* 357, 1420.
Neumayer, E. (2005): Bogus Refugees? The determinants of asylum migration to Western Europe. *International Studies Quarterly* 49(3), 389–410.
Perch-Nielsen, S. (2004): Understanding the effect of climate change on human migration – The contribution of mathematical and conceptual models. *Diploma Thesis – Department of Environmental Sciences ETH – Zurich.*
Perch-Nielsen, S., M.B. Bättig, and D. Imboden. (2008): Exploring the link between climate change and migration. *Climatic Change* vol. in print.
Renaud, F., J.J. Bogardi, O. Dun, and K. Warner. (2007): Control, adapt or flee how to face environmental migration? *InterSecTions 'Interdisciplinary Security ConnecTions' Publication Series of UNU-EHS* 5.
Reuveny, R. (2007): Climate change-induced migration and violent conflict. *Political Geography* 26(6), 656–673.
Reuveny, R. (2008): Ecomigration and violent conflict: Case studies and public policy implications contact information. *Human Ecology* 36(1), 1–13.
Richmond, A.H. (1988): Sociological theories of international migration: The Case of Refugees. *Current Sociology* 36(2), 7–25.
Richter, R.E. (1998): Umweltflüchtlinge in Afrika. In: Scheffran, J., and Vogt, W.R. (ed.), *Kampf um die Natur*. Darmstadt. pp. 42–73.
Robinson, A., J. Lindbergh, and L. Brinkman. (1961): A correlation and regression analysis applied to rural farm population densities in the Great Plains. *Annals of the Association of American Geographers* 51, 211–221.
Sen, A.K. (1981): *Poverty and Famines: An Essay on Entitlement and Deprivation*. Clarendon Press, Oxford.
Smith, K. (2001): *Environmental Hazards, Assessing the Risk and Reducing Disaster*. Routledge, London.
Stern, N. (2006): *The Stern Review on the Economic Effects of Climate Change (Report to the British Government)*. HM Treasury – Retrieved from www.sternreview.org.uk
Thill, S. (2008): Climate destruction will produce millions of 'Envirogees'. *Alternet*. http://www.alternet.org/environment/86285/?page=1 – visited 27th June 2008. no. May 27.

UNHCR. (2008): *Climate change, environmental degradation, natural disasters and human displacement: UNHCR's preliminary observations (second draft)*. The Office of the United Nations High Commissioner for Refugees (mimeo), Geneva.

United Nations University. (2005): As ranks of "Environmental Refugees" swell worldwide, calls grow for better definition, recognition, support. *11 October 2005 – Press Release – UNU Institute for Environment and Human Security* no. Retrieved on 27 July, 2007. Available from http://www.ehs.unu.edu/article:130.

Westing, A.H. (1992): Environmental refugees: A growing category of displaced persons. *Environmental Conservation* 19(3), 201–207.

Wood, W.B. (2001): Ecomigration: Linkages between environmental change and migration. In: Zolberg, A., and Benda, P.M. (ed.), *Global Migrants Global Refugees – Problems and Solutions*. Berghahn, New York, NY. pp. 42–61.

Zolberg, A.R., A. Suhrke, and S. Arguayo. (1986): International factors in the formation of refugee movements. *International Migration Review* 20(2), 151–169.

Chapter 7
Forced Migration of Alaskan Indigenous Communities Due to Climate Change

Robin Bronen

1 Introduction

In the Arctic, climate change is evident and occurring at a rate faster than other parts of the planet. The 2007 physical science report of the Intergovernmental Panel on Climate Change (IPCC, 2007) confirms that temperatures in the Arctic have increased by approximately twice the global average between 1965 and 2005 (IPCC, 2007: 339). Since 1975, temperatures in Alaska have increased by an average of 2–3.5°C. In addition, Arctic sea ice is decreasing in extent and thickness, wildfires are increasing in size and frequency, and permafrost is thawing. These phenomena are creating a humanitarian crisis for the indigenous communities that have inhabited the arctic and boreal forest for millennia. Approximately 200 indigenous villages are located along the navigable waters of Alaska's coasts and rivers. Dozens of these communities are threatened because of accelerated rates of erosion or flooding due to climate change. Governments are struggling to respond. The traditional adaptation strategies of hazard prevention and disaster relief are no longer protecting communities. Relocation of the communities is required.

This chapter adds to the body of research focused on defining the nexus between climate change and human migration. The empirical research discussed in this chapter identifies a migration type, permanent relocation, which specifically links climate change with human migration and demonstrates that new institutions need to be created in order to protect the human rights of those forced to migrate and prevent humanitarian crises. An accurate definition of this displacement category is essential in order to ensure that the permanent relocation of communities only occurs when there are no other durable solutions. A precise definition is also critical to the design and implementation of institutional frameworks of humanitarian response. This institutional framework needs to be based in human rights doctrine and created at the local, national, and international levels.

R. Bronen (✉)
Resilience and Adaptation Program, University of Alaska, Fairbanks, AK, USA
e-mail: rbronen@yahoo.com

T. Afifi, J. Jäger (eds.), *Environment, Forced Migration and Social Vulnerability*,
DOI 10.1007/978-3-642-12416-7_7,

2 Identifying the Nexus Between Climate Change and Human Migration

Climate change will create disparate environmental impacts that force people and communities to migrate. Erosion, flooding, and sea-level rise will be the primary causes of displacement. Water and food security issues, due to drought and salt water intrusion, will also impact the sustainability of communities and cause migration.

The disparate drivers of climate-induced migration can be segregated into three distinct categories: random extreme weather events, such as hurricanes and tornadoes, the depletion of ecosystem services, such as drought and salt water intrusion, and on-going ecological changes caused by the combination of random extreme weather events and depletion of ecosystem services that severely impact public infrastructure, such as health clinics and schools, as well as the livelihoods and lives of the people residing in the community.

These climate change drivers cause distinct patterns of human migration. The three migration typologies are the migration of individuals and households where climate change is one of several factors causing migration, mass migration where entire communities are forced to temporarily evacuate, and mass migration where entire communities are forced to permanently relocate. Each migration typology requires a distinct institutional adaptation strategy to ensure that the humanitarian response is appropriate and that people's human rights are protected.

Research has primarily focused on the first migration typology and has documented the difficulty in demonstrating the nexus between climate change and migration. These studies primarily focus on the depletion of ecosystem services as the primary climate change factor that causes the migration of individuals and households. Some of the studies do not distinguish between ecosystem depletion caused by human overuse of resources and ecosystem depletion caused by climatic events, such as drought (Hugo, 1996; Morales, 2008). This distinction is critical when implementing institutional adaptation strategies.

Socio-economic factors create the difficulty in isolating climate or environment as a significant or exclusive factor that causes migration. In Leighton's (2008) survey of desertification and drought-related migration, population growth, household income, social networks, and employment opportunities intertwine with climate factors as the causes for migration. Migration is seen as one of many coping strategies to alleviate poverty.

In the recent International Organization for Migration (IOM) publication, "Climate Change and Migration: Improving Methodologies to Estimate Flows," the authors review several case studies which analyze the connection between drought and migration. These studies focus on the consequences of ecosystem depletion on livelihood decisions and primarily analyze the decision-making process of individuals and households to assess the likelihood of migration. The research affirms that a multiplicity of socio-economic factors, along with climate change, push people to migrate (Kniveton, Schmidt-Verkerk, Smith and Black, 2008).

Similarly, Hugo (1996) argues that climatic causes of migration are difficult to segregate from the "pre-disposing" socio-economic factors such as population density, poverty, and resource use (Hugo, 1996: 109). Although his research did not distinguish between ecosystem degradation caused by human overuse of resources and ecosystem depletion caused by climate change, his analysis reveals the difficulty in defining the nexus between migration and climate change when ecosystem depletion is the climate change factor driving the migration.

The IOM climate change and migration study also examines the mass population displacement caused by random environmental events, where communities are temporarily evacuated and infrastructure is rebuilt with the expectation that people will return home. In this context as well, research has focused on the migration patterns of individuals or households who chose not to return home. In some of these studies, researchers identified socio-economic status as a factor that contributes to the migration decision and demonstrated again the difficulty of segregating climate from other reasons that cause people to migrate (Kniveton, Schmidt-Verkerk, Smith and Black, 2008).

The focus of this chapter is on the third migration type, permanent relocation caused by an on-going depletion of ecosystem services created by extreme weather events or climate-induced change, such as sea-level rise. *Climigration* is the word that best describes this type of human migration. *Climigration* occurs when a community is no longer sustainable exclusively because of climate-related events and permanent relocation is required to protect people. The critical elements of this migration type are that climatic events are on-going and repeatedly impact public infrastructure and threaten people's safety so that loss of life is possible. This type of human migration is occurring in Alaska. Catastrophic random environmental events, such as tornadoes and hurricanes, do not cause *climigration*. However, these random environmental events, if on-going, may alter ecosystem services permanently, cause extensive damage to public infrastructure, repeatedly place people in danger, and require communities to relocate.

3 Climate-Induced Migration in Alaska

Erosion and flooding have plagued the hundreds of indigenous villages that are located along the coast and rivers of Alaska for decades. These communities rely on easy access to navigable waters to hunt marine mammals and fish in order to subsist. Subsistence is central to their culture and survival. Arctic sea ice and frozen tundra are essential ecosystem services. Arctic sea ice has, in the past, protected communities from coastal erosion and flooding by creating a barrier to storm-related waves and surges. Along the northwestern Alaskan coast permafrost, permanently frozen subsoil, is continuous and the glue that keeps the land intact and habitable.

Climate change is depleting these ecosystem services and accelerating the rates of erosion. The complex interplay of ecological feedback loops is now endangering the lives of the inhabitants of these communities. As Mark Serreze, senior snow and ice scientist in Boulder Colorado, stated in a newspaper interview in December 2007, "The Arctic is screaming." (Borenstein, 2007). Temperatures along the northern Alaskan coast have increased by an average of 3.5°C during the winter since 1975 (Shulski and Wendler, 2007: 134). These warming temperatures are causing the permafrost to thaw and the Bering and Chukchi Seas to freeze later in the autumn. The 2007 IPCC physical science report has documented that the temperature of the top layer of permafrost has increased by up to 3°C since the 1980s (IPCC, 2007: 339).

At the same time, arctic sea ice is decreasing in thickness and extent. Recent research concluded that the North Pole had a 50% chance of being ice-free during the summer of 2008 (Borenstein, 2008). Record minimum levels of arctic sea ice have been recorded since 2002. The decrease in extent of arctic sea ice coupled with warming temperatures has caused a delay in freezing of the Bering and Chukchi Seas. Nearshore pack ice has historically provided a protective barrier to coastal communities. Research has documented that since the 1980s the arctic seas are remaining ice-free approximately 2 months longer in the autumn (US Army Corps of Engineers, 2006). The delay in freezing of the arctic seas has left many communities exposed to the autumnal storms that originate in the Pacific and occur primarily between August and early December. Bering Sea storms are not hurricanes but do have hurricane strength in the damage they can cause on the coast due to wave action and storm surges (Atkinson, 2007). The loss of arctic sea ice coupled with thawing permafrost is causing severe erosion and storm surges that are threatening the lives of the inhabitants of several communities.

Five indigenous communities, located along the western outer coast of Alaska on the Bering and Chukchi Seas, have concluded that relocation is the only durable solution to the climatic events that are threatening their lives. These communities are witnessing dramatic climate shifts that are impacting community infrastructure as well as the safety of the people who reside within each community. Government agencies recognize that each of these communities is imminently threatened by erosion and flooding and are struggling to meet the enormous new needs of these communities. Up until 2006, local, state, and federal government agencies have responded within the context of disaster relief. Millions of dollars have been spent on shoreline protection and consulting firms to study the issue of erosion and relocation. Government agencies now realize that erosion control, the traditional disaster relief response, is a short-term solution and no longer able to protect the inhabitants of these coastal communities. In 2006, a US government report found that a catastrophic climatic event could submerge three of these communities within 10–15 years of the report's publication (US Army Corps of Engineers, 2006).

This chapter focuses on three communities identified in the 2006 US government report. The tribal governments of Newtok, Shishmaref, and Kivalina recognized several years ago that relocation was the only durable solution that would protect their respective communities. Each community has been engaged for years in an ad hoc process to secure state and federal government assistance in their relocation efforts. The residents of each community have voted and decided that their preferred relocation alternative is to move to a new site and recreate their community. Each community has commissioned several studies to engage in a socio-ecological assessment and evaluate the relocation process. All studies demonstrate the complexity of the issues each community faces, if relocation occurs. The cultural, social, economic, and psychological impact of the displacement is incalculable.

3.1 Newtok

Newtok is a traditional Yu'pik Eskimo village located along the Ninglick River near the Bering Sea. The village site is within the Yukon-Kuskokwim Delta, one of the largest river deltas in the world, and surrounded by marshy tundra and thousands of lakes (Cox, 2007). The people of Newtok are known as "Qaluyaarmiut" or "dip net people" whose ancestors have inhabited the Bering Sea coast for at least 2,000 years (ASCG, 2004). According to the 2000 Census, 321 people reside in the community (ASCG, 2004). The only access to the community is by barge during the summer or by airplane.

Newtok has made the most progress in its relocation efforts; in 2003, the Newtok Native Corporation secured land ownership to a relocation site through a legislative act of the US Congress. The new community is located 9 miles south of Newtok across the Ninglick River and has been named Mertarvik, which in Yu'pik means "getting water from the spring" (Cox, 2007).

The change of the Ninglick River course is causing severe erosion of the river bank adjacent to the village. Climate change is impacting the rate of erosion, which is caused by a combination of increased temperatures, thawing permafrost, wave action, and river current (Cox, 2007). Erosion, flooding, and salt water intrusion are threatening the community. Critical public infrastructure is being washed away (Cox, 2007: 8–11). The barge landing, which provides summer access to the community for supplies and fuel for heating, no longer exists and is causing a fuel crisis. Salt water is affecting the potable water. Community inhabitants are experiencing mental and physical health issues associated with the decline of ecosystem services, such as potable water (Cox, 2007: 8).

The community has monitored the erosion rates of the Ninglick River for decades. In 1994, the Newtok Traditional Council started a relocation planning process and analyzed relocation to six potential village relocation sites. The

Newtok Traditional Council also considered relocation to three already existing villages with residents of Newtok being dispersed between them (ASCG, 2004). Ten years later, the Newtok Traditional Council commissioned a report to provide background documentation to government agencies and officials to justify the efforts of the village to relocate and to support requests for government assistance in this process (ASCG, 2004).

In 2006, an Alaska state agency created the Newtok Planning Group to specifically address the short-term emergency needs of the community and to begin a relocation planning effort. Approximately 25 different tribal, state, and federal government representatives participate in the Newtok Planning Group (Cox, 2007: 13). The state agency coordinates their work, but has no dedicated funding to relocate the community and no jurisdictional authority to require other agencies to perform work needed for the relocation. For these reasons, the Newtok Planning Group has encountered numerous hurdles that have slowed their progress.

The Newtok Traditional Council built three homes at their new village site, Mertarvik, in September 2006. The Alaska Department of Transportation and Public Facilities, a member of the Newtok Planning Group, plans to build the barge landing at Mertarvik during the summer of 2009 (Cox, 2007: 13–18). The tribal government believes that their village must be relocated by 2012 to avoid a greater humanitarian crisis.

3.2 Kivalina

Kivalina is an Inupiaq Eskimo village located on a barrier island in the Chukchi Sea, approximately 80 miles above the Arctic Circle. People have lived at this location for at least 150 years. Approximately 360 people currently live in Kivalina (Swan, 2007a). Subsistence is the lifeblood of the community. The Native Village of Kivalina, a federally recognized indigenous tribe, has been working on its village's relocation for several decades (Swan, 2007b). Erosion has historically been a concern for the community. In 1963, the Native Village of Kivalina voted on relocation, but a majority of residents did not want to move, so the community remained in its present location (Swan, 2007a). In 1998, the community voted a second time on the relocation issue and decided to relocate. The community chose a relocation site (Swan, 2007a). Federal government agencies later determined that the chosen site was unsuitable for community development because of permafrost (Tryck, 2006). The community voted on relocation a third time in 2000 and chose a different relocation site (Swan, 2007a). Despite this 2000 vote, the community has not begun to relocate, because federal government agencies believe that the second site chosen by the community is also unsuitable for relocation (Swan, 2007a).

At the present time, erosion is causing multiple harms to the community, including loss of critical public infrastructure and transportation access to the community. Air and sea are the only two methods of access to the community. The silt build-up from erosion is affecting summer barge access to Kivalina, and erosion caused by sea storm surges is impacting the community's airstrip. In addition, private homes have had to be relocated into the interior of the barrier island (Tryck, 2006: 6). Erosion is also threatening the stability of the solid waste storage containment area on the island, which could create an environmental catastrophe for the surrounding bodies of water (Tryck, 2006: 5–6).

Erosion control continues to be the primary focus of government assistance. In September 2006, federal government leaders arrived in Kivalina to celebrate the finalization of a multi-million dollar seawall. Prior to the commencement of celebrations, a storm damaged 160 feet of a 1,800 foot seawall and caused the officials to cancel the celebration (deMarban, 2006). One year later, in September 2007, a storm once again threatened the community, and its residents feared that the seawall would not provide them with protection; 250 Kivalina residents evacuated their community in search of safety (Bragg, 2007).

Recently state funding has been allocated to start a relocation planning process. The community believes that they must be relocated within the next 5 years to prevent a greater humanitarian crisis (Black and Opheen, 2008).

3.3 Shishmaref

Shishmaref is a traditional Inupiat Eskimo community located on Sarichef Island on the northwest coast of Alaska. The barrier island separates the Chukchi Sea from a saltwater lagoon on the leeward side of the island. The residents of Shishmaref have lived on the island for several centuries and actively subsist (Weyiouanna, 2007). Access to the ocean is essential to their survival. For several decades, village residents have been concerned about ongoing erosion of their island. Historically, state and federal government agencies have responded by investing in shoreline protection, which has provided a barrier for the community (TetraTech and Inc, 2004).

In the last decade, autumn storms have increased in their intensity, and storm surges from the Chukchi Sea have endangered their community. During the same time period, residents have also documented that the freeze-up of the Chukchi Sea is occurring more than 3 weeks later, in mid-November instead of late October (Hufford and Partain, 2005). In 2002, residents documented that the sea did not freeze until the end of December (Hufford and Partain, 2005). Since 2002, Shishmaref has witnessed increasing winter temperatures coupled with thawing permafrost and intense autumn storms. In February 2006, the community documented open water in the Chukchi Sea, an event that had never been witnessed by any resident of their community prior to that time (Weyiouanna, 2007).

Erosion control is no longer providing protection during the autumnal storm season. Most historical records point to October 1997 as the beginning of the current erosion crisis. At that time, an autumn storm caused severe erosion and required 14 homes and the National Guard Armory to be relocated (TetraTech and Inc, 2004; US Army Corps of Engineers, 2006). Severe autumn storms since 2002 have threatened homes, power facilities, and other critical village infrastructure (TetraTech and Inc, 2004).

In 2002, the Native Village of Shishmaref created a relocation and erosion coalition (Weyiouanna, 2007). In July 2002, residents voted to relocate the community (Weyiouanna, 2007). Several federal government agencies have studied the erosion and relocation issue since 2002, but the actual relocation of the community has not yet started (TetraTech and Inc, 2004). The tribal government of Shishmaref believes that their community must relocate by 2012 (Weyiouanna, 2007).

4 Traditional Governance Structures Will Not Protect Communities

In Alaska, government agencies have slowly recognized that the traditional adaptation strategies of erosion control and flood relief will not protect communities and that the traditional governance structures that provide disaster relief are not adequate to respond to the complex humanitarian emergencies facing Alaskan communities. Several reports have been published since 2003, documenting the enormity of the problem. In addition, since the summer of 2007, government officials have organized numerous public hearings to gather testimony and to begin to implement a new governance structure.

In 2003, the US Congress authorized a federal agency to analyze the extent of the erosion problem affecting Alaskan communities (United States General Accounting Office, 2003). The report documented that erosion and flooding are affecting 184 of 213 communities, 9 of these communities were physically threatened and 4 of these 9 communities are in imminent danger of loss of life and property (United States General Accounting Office, 2003). Shishmaref, Kivalina, and Newtok were three of the communities listed that are in imminent danger because of erosion and flooding.

Subsequently, Congressional legislation authorized a second report to conduct a more comprehensive assessment of the 9 communities identified in the 2003 report. The 2006 Alaska Village Erosion Technical Assistance Program report evaluated the different costs associated with erosion control versus relocation (US Army Corps of Engineers, 2006). Although the report primarily focused on the different costs associated with each adaptation strategy, the report also identified a number of critical governance issues that need to be addressed, if relocation occurs. The report recognized that no government agency has the authority to relocate communities and that no funding is specifically designated for relocation (US Army Corps of Engineers, 2006). In addition, no criteria exist to choose a relocation site. Most importantly, the report recognized that no governmental organization exists that can address the strategic planning needs of relocation and the logistics of decommissioning the original community location, including hazardous waste clean-up and preservation of cultural sites (US Army Corps of Engineers, 2006).

In 2007, the Governor of Alaska created the Alaska Climate Change Sub-Cabinet to implement a climate change strategy for the state. The Immediate Action Workgroup is one of the advisory groups to the Sub-Cabinet and was tasked with identifying the short-term emergency steps that state government must take to prevent loss of life and property due to climate change in the communities that must relocate. A state and federal government representative are co-chairs of the Workgroup (Black and Opheen, 2008). The multi-level governance structure of the advisory group is unique. In April 2008, the Workgroup issued its recommendations to the Governor and made numerous recommendations to address the immediate humanitarian needs of the communities forced to relocate (Black and Opheen, 2008).

Erosion control and community evacuation plans are central to the recommendations. The Workgroup also recommended that funding be allocated to communities to begin a relocation planning process. The Workgroup recognized the complex governance issues identified in the 2006 Alaska Village Erosion Technical Assistance Program report and recommended that one state agency leads the relocation effort and acts as the coordinating agency with responsibility of maintaining federal, state, and tribal partnerships (Black and Opheen, 2008). The report, however, does not detail the governance structure or jurisdictional authority that will allow the agencies to work together.

The Newtok Planning Group continues to be the only multi-level interdisciplinary governmental workgroup in Alaska focused on relocation. The state planner facilitating the workgroup is coordinating the work of the dozens of agencies involved with Newtok's relocation. She has no jurisdiction to require other agencies to join in her relocation efforts, but federal and state agencies are working with the Newtok Traditional Council and willingly engaging in the relocation process. The Newtok Traditional Council is a small local tribal government that has limited capacity to coordinate the relocation work of dozens of federal and state agencies and administer and obtain funding needed for the relocation process.

The unanswered question is whether the organizational structure of the Newtok Planning Group will be replicated for the other communities that need to relocate without specific legislative authority. None of the agencies participating in the Newtok Planning Group have a funded mandate to relocate communities endangered by climate change. Several agencies are bound by legal guidelines that have created barriers to Newtok's relocation effort. State agencies, such as the Alaska Department of Transportation designated with the task of building airstrips and the Alaska Department of Education designated with building schools, are unable to move forward with these projects at the relocation sites because the regulations require that an existing community with a minimum population be at the site where this infrastructure is built.

In addition, there is no dedicated funding source for relocation and no lead agency designated with creating a statewide relocation strategic plan and coordinating the various agencies working on housing, transportation, community infrastructure, education, health, and related socio-economic needs. There is also no strategic plan for cleaning-up and securing the old village sites, so that they can be abandoned without creating an environmental hazard. These governance issues must be addressed and resolved in order to ensure that the communities being forced to migrate because of climate change are protected.

5 Creating New Governance Structures

Climigration requires new and unique institutional responses. Communities, rather than individuals or households, will be forced to permanently migrate. Permanent relocation will be mandated, because there will be no ability to return home, since homes will be under water or sinking in thawing permafrost.

The relocation of communities will require new multi-level interdisciplinary governance structures to address the complexity of issues that arise when a community is forced to migrate due to climate change. Human rights doctrine must guide the creation of new institutional frameworks. Refugee law, the Universal Declaration of Human Rights, the Guiding Principles on Internal Displacement, and the recently adopted Universal Declaration on the Rights of Indigenous Peoples provide a theoretical basis for creating these principles. However, none of these legal documents address the complex crises of populations facing *climigration*. International human rights principles need to be specifically created for *climigration* to ensure that the human rights of those forced to migrate because of climate change are protected.

In addition, the work of the United Nations Office for the Coordination of Humanitarian Affairs (UN OCHA) and the World Bank's Involuntary Resettlement policies and procedures can provide operational guidance to governments faced with relocations caused by climate change. The UN OCHA office coordinates multi-level governance structures to provide humanitarian assistance in complex emergencies. OCHA recognizes that there are two different types of field coordination mechanism, strategic and operational (OCHA, 1999: 22). Strategic coordination involves the creation of the overall plan to guide the humanitarian relief effort, the allocation of responsibilities among the disparate international, national, and local agencies, monitoring the humanitarian relief effort, and mobilizing and allocating resources (OCHA, 1999: 22). Operational coordination involves the coordination of specific sectors of the humanitarian assistance. This type of cross-scale interdisciplinary governmental coordination will be essential when responding to the needs of communities forced to migrate because of climate change.

The World Bank has expertise in the resettlement of communities when governments undertake development projects (World Bank, 2004). The procedures developed by the World Bank in response to the social, cultural, economic, and political challenges of resettlement also provide important information to communities and governments faced with relocation due to climate change. The guidance provided by the World Bank will be critical in order to avoid the impoverishment and social disintegration that has historically plagued the involuntary resettlement of communities because of government-sponsored development projects.

6 Conclusion

Climate change is forcing communities to migrate. Human migration because of climate change will present one of the most severe challenges to governments tasked with providing humanitarian assistance and protection. Alaskan indigenous communities are at the forefront of this global humanitarian crisis. Traditional governance structures will be inadequate to respond to the needs of those forced to migrate. The international community must create guidelines, based in human rights doctrine, for national and local governments to ensure that the human rights of those forced to migrate because of climate change are protected.

References

Alaska Community Database Summaries, Shishmaref. Available from http://www.commerce.state.ak.us/dca/commdb/CIS.cfm?Comm_Boro_Name=Shishmaref

Alaska Community Database Summaries, Kivalina. Available from http://www.commerce.state.ak.us/dca/commdb/CIS.cfm?Comm_Boro_Name=Kivalina

ASCG Incorporated. (2004): *Newtok Background for Relocation Report*. ASCG Incorporated, Anchorage, AK.

Atkinson, D. (2007): *Coastal Hazards in Alaska: Threats, Trends and Needs*. Presentation at the University of Alaska, Fairbanks. 6 November.

Black, M., and P. Opheen. (2008): *Recommendations Report to the Governor's Subcabinet on Climate Change*. Immediate Action Workgroup, Anchorage, AK.

Borenstein, S. (2008): Odds increase for ice-free North Pole this summer. *Anchorage Daily News*. 28 June. p. A1.

Borenstein, S. (2007): As ice melts, so do records. *Anchorage Daily News*. 12 December. p. A1.

Bragg, B. (2007): Fierce fall storm pounds Kivalina after most villagers flee. *Anchorage Daily News*. 14 September. p. A1.

Cox, S. (2007): *An Overview of Erosion, Flooding, and Relocation Efforts in the Native Village of Newtok*. Alaska Department of Commerce, Community and Economic Development, Anchorage, AK.

deMarban, A. (2006): New wall takes sea's first test. *Anchorage Daily News*. 15 September. p. B1.

Hufford, G., and J. Partain. (2005): *Climate Change and Short-Term Forecasting for Alaskan Northern Coasts*. National Weather Service, Anchorage, AK.

Hugo, G. (1996): Environmental concerns and international migration. In: *International Migration Review*. The Center for Migration Studies, New York, NY.

International Organization for Migration. (2007): Migration and the environment, ninety-fourth session. Discussion Note. International Organization for Migration, Geneva.
IPCC. (2007): *Climate Change 2007: The Physical Scientific Basis*. Contributions of Working Group I to the Fourth Assessment Report of the Intergovernmental Panel on Climate Change. In: Solomon, S., Qin, D., Manning, M., Chen, Z., Marquis, M., Marquis, K.B., Averyt, K.B., Tignor, M., and Miller, H.L. (eds.), Cambridge University Press, Cambridge, UK and New York, NY.
Kniveton, D., K. Schmidt-Verkerk, C. Smith, and R. Black. (2008): *Climate Change and Migration: Improving Methodologies to Estimate Flows*. International Organization for Migration, Geneva.
Leighton, M. (2008): Summary of desertification and drought related migration. *Paper for the 16–18 April Expert Workshop on Environment and Migration*, Munich.
Morales, C. (2008): Desertification, degradation and migration in Latin America and the Caribbean. *Paper for the 16–18 April Expert Workshop on Environment and Migration*, Munich.
OCHA. (1999): *OCHA Orientation Handbook on Complex Emergencies*. OCHA, New York, NY.
Shulski, M., and G. Wendler. (2007): *The Climate of Alaska*. University of Alaska Press, Fairbanks, AK.
Swan, C. (2007a): A presentation by the native village of Kivalina and the City of Kivalina to the Alaska climate impact assessment commission, Anchorage, AK, June 28.
Swan, C. (2007b): A presentation to the subcommittee on disaster recovery HSGAC, Anchorage, AK, October 11.
TetraTech, Inc. (2004): *Shishmaref Partnership Shishmaref Relocation and Collocation Study*. TetraTech, Inc., Anchorage, AK.
Tryck, N.M. (2006): *Kivalina, Alaska: Relocation Planning Project Master Plan*. U.S. Army Corps of Engineers, Anchorage, AK.
US Army Corps of Engineers. (2006): *Alaska Village Erosion Technical Assistance Program*. US Army Corps of Engineers, Anchorage, AK.
United States General Accounting Office. (2003): *Alaska Native Villages: Most Are Affected by Flooding and Erosion, but Few Qualify for Federal Assistance*. General Accounting Office, Washington, DC.
Weyiouanna, T., Sr. (2007): Shishmaref erosion and relocation coalition concerns. *Immediate Action Workgroup Hearing*. Fairbanks, AK, November 6.
World Bank. (2004): *Involuntary Resettlement Sourcebook: Planning and Implementation in Development Projects*. World Bank, Washington, DC.

Chapter 8
'Buscando la vida' – How Do Perceptions of Increasingly Dry Weather Affect Migratory Behaviour in Zacatecas, Mexico?

Kerstin Schmidt-Verkerk

1 Introduction

'Buscando la vida' – looking for life – is a phrase that many people in the Mexican state of Zacatecas refer to when they talk about their motives for migration. Life in the rural communities of Zacatecas is to a large extent determined by farming and migration. Due to a lack of industries and a decline in the mining activities in Zacatecas, agriculture and remittances are the major sources of income for many people. As Delgado Wise et al. (2004: 168) calculated, based on data of the 2000 census, in Zacatecas, 61.5% of the households indicate that remittances are the most important source of household income and 34.8% indicate that remittances are the only source of household income. Agriculture exists in two forms: (1) subsistence agriculture on the 'ejido' [1] land, which is usually the rain-fed cultivation of maize and beans and (2) commercial agriculture, which often makes use of irrigation systems or even greenhouses. People from many rural communities work seasonally, as day labourers in commercial production.

In recent years, changing precipitation patterns in combination with increasing prices for seeds, fertilisers and agricultural machines have been affecting subsistence and commercial agriculture and therefore people's livelihoods. Rising prices of consumer goods as well as low wages have contributed to the difficult economic situation of the majority of village dwellers. This chapter examines how people respond to these challenges to their livelihood and looks into the question of if – and to what extent – these challenges affect different forms of migration that people have been using as a livelihood strategy for many decades.

K. Schmidt-Verkerk (✉)
Sussex Centre for Migration Research, University of Sussex, Brighton, UK
e-mail: k.schmidt-verkerk@sussex.ac.uk

[1] An 'ejido' is communally owned land that was given to the people with the fall of the Hacienda system after the revolution of 1910. People in some communities cultivate the land together, while in most communities all families received a share when the land was split up. Nowadays many families do not have rights on the 'ejido' land because they moved into the community later, they or their parents already sold the land or their parents who own the land are still alive.

T. Afifi, J. Jäger (eds.), *Environment, Forced Migration and Social Vulnerability*, 99
DOI 10.1007/978-3-642-12416-7_8, © Springer-Verlag Berlin Heidelberg 2010

2 Theoretical Background: The Climate Change–Migration Link

The potential effects of environmental change on migration have been explored theoretically for more than three decades. Building on the arguments in this debate, the nexus between climate change and migration has become topical only very recently. The assumption that environmental change – and consequently climate change – will lead to international mass migration has been contested theoretically and has also been challenged by empirical research.

The concept 'environmental refugee' was introduced by Lester Brown of the Worldwatch Institute in the 1970s (Saunders, 2000: 218–219), but became more widely known through the work of El-Hinnawi (1985) of the United Nations Environment Programme (UNEP) a decade later. In 1995, Norman Myers and Jennifer Kent proposed the first estimates of numbers of current and future 'environmental refugees'. However, Jodi Jacobsen in 1988 already concluded that 'environmental refugees have become the single largest class of displaced persons in the world' (Lonergan and Swain, 1998). Other scholars (Black, 1998; 2001; Kibreab, 1997; McGregor, 1993; Suhrke, 1994) have challenged the concept 'environmental refugee' and the estimation of numbers of people who are expected to become displaced or migrate due to environmental problems for being oversimplistic. Numbers of people who are projected to become displaced are based on population estimates in certain regions, as reliable statistics rarely exist in many of these areas. Furthermore one cannot assume that the whole population will leave after some form of change to the environment has occurred. It is even more unlikely that everybody who does leave a district will cross an international border or even cross continents or seas to reach an economically more developed Western country. There is broad theoretical consensus that it is generally not the poorest people who migrate overseas, because international migration is an expensive endeavour that demands resources for the journey and for the crossing of borders (Castles, 2000: 3; de Haan, 2000: 7; Skeldon, 2002: 71).[2] It is thus difficult to imagine that people who have lost their land and all their other resources will manage to embark on a journey to Europe or to North America. This criticism is supported by the general consensus that migration is a very complex phenomenon which cannot be explained by one single reason alone (Boyle et al., 1998; Castles and Miller, 1993; Kritz et al., 1992; Wood, 2001). These authors argue that, on the contrary, there is a multitude of economic, social and cultural reasons that influence migratory behaviour.

The nexus between weather phenomena that are attributed to climate change and migration has gained growing attention after the release of the *Summary for Policymakers of the Working Group II document on Impacts, Adaptation, and Vulnerability*, which is part of the Fourth Assessment Report

[2] Most authors acknowledge that in exceptional cases, the poorest might become displaced because of environmental catastrophes or wars. However, they are not specific about the destinations of these people.

of the Intergovernmental Panel on Climate Change. This report mentions the 'potential for population migration' due to increases in the number of areas affected by droughts and an increase in intense tropical cyclone activities (IPCC, 2007: 16). Despite this rather cautious statement, the expression of a link between climate change and migration has led to considerable coverage of the subject in newspaper and magazine articles and in the policy papers of different NGOs. Estimates of the numbers of expected 'climate change refugees' draw on figures put forward by Myers and Kent (1995) in the context of the 'environmental refugee' debate. These figures of 150–200 million climate change refugees in the coming 30 years are quoted in a report written on behalf of Greenpeace Germany (Jakobeit and Methmann, 2007). According to the *Stern Review* on the economics of climate change (2006: 77), these numbers are based on 'conservative assumptions'. The report goes on to argue that 'climate change will lead to hundreds of millions more people without sufficient water or food to survive and/or threatened by dangerous floods and increased disease' (Stern, 2006: 77). Accordingly, a report for the British organisation Christian Aid (2007: 22) states that there will be 1 billion climate change refugees by 2050. However, the authors admit that as not enough research has been carried out so far, there are no recent, authoritative, global figures on the number of people who could be displaced from their homes by climate change (Christian Aid, 2007: 22).

2.1 Empirical Results So Far – How Does Climate Variability Influence Migratory Behaviour?

The scarcity of empirical studies of the nexus between climate change and migration might partially be caused by the methodological problem that arises due to the fact that climate change is an abstract concept, most of the consequences of which will be felt in the future. All that empirical studies can directly observe is how current climate variability[3] affects migration. Changes in rainfall patterns or drought are one form of this climate variability. The following paragraphs summarise the results of a selection of these studies.

One of the first studies that examine the relationship between drought and migration is Sally Findley's (1994) research on migration from rural Mali during the 1983–1985 drought. She found that long-distance migration – notably of male household members – to France decreased. This can be explained by the fact that food scarcity – leading to increased prices – forces people to spend more money on their basic needs. Therefore, they cannot afford to invest in migration anymore. At the same time, short-distance migration to larger agglomerations increased, because women and children were sent in search of work to contribute to household incomes. In addition, this strategy reduced the

[3] See UNFCCC (1992: 3) for definitions of 'climate change' and 'climate variability'.

number of persons in a household, and thus, the amount of food needed. Under the comparable circumstances of the 1980s drought in the Sahel, Ezra and Kiros (2001) analysed rural out-migration in the drought-prone areas of Ethiopia. Almost 80% of the people moved to other rural areas. The authors do not specify whether these moves involved border crossings or not, but it is not likely that long distances were covered, as the destination of a long-distance move is usually a bigger agglomeration. Contrary to their expectations, Ezra and Kiros found that the percentage of people who gave reasons for moving that might be related to drought[4] was very low. Most migration was caused by family formation after marriage. A multilevel analysis of their survey data, however, revealed that, next to age and gender, the availability of food at the community level was identified as a major factor that determined out-migration. Ruth Haug (2002) welcomes the classification of pastoralists in Northern Sudan, whose livelihoods became affected by the mid-1980s drought, as 'environmental refugees'. She thinks that many of them would have starved if they had not migrated. Nevertheless, she acknowledges that migration always has been a livelihood strategy in the Hawaweer society and that some of them were forced to stay during the drought, because they did not have the resources to migrate (Haug, 2002: 76). Haug thus stresses that the consequences drought has upon livelihood decisions, including migration, largely depend on the socioeconomic situation of the people concerned. Accordingly, Elisabeth Meze-Hausken (2004) points out that vulnerability to drought alone does not cause migration. In her study in Northern Ethiopia, in which she surveyed more than 100 peasants, she found that 'people in marginal regions have developed a great variety of adaptation mechanisms, which strengthen their ability to cope with both, slow climatic changes and extreme climatic events' (Meze-Hausken, 2004: abstract). Henry, Schoumaker and Beauchemin (2004) investigated the effect of changing rainfall patterns on migration in Burkina Faso by using event history analysis. They found no relationship between changing rainfall patterns and migration in general when they did not distinguish between different types of migration by destination and duration. Individual characteristics of people, such as level of education, type of activity involved in, and belonging to a particular ethnic group, seemed to be the deciding factors for migration. Once split up between different types of migration, the study reveals that people living in areas with scarce rainfall are much more likely to engage in short-distance moves than people living in other regions. However, the number of migrants does not increase after periods of minimum rainfall in the dryer regions. The fundamental conclusion that the authors draw from their findings is that: 'long-term migrations seem to be less related to environmental conditions than short-term moves' (Henry et al., 2004: 455).

The empirical results discussed above reveal that the nexus between drought and migration is not straightforward. The five studies discussed produced two

[4] These reasons are 'drought', 'shortage of land' and 'searching for work'.

general results; first, drought seems to cause an increase in the number of people who engage in short-term rural to rural types of migration. On the other hand, it does not affect, or even decrease international, long-distance moves. Second, the conceptualisation of vulnerable people as helpless victims who have to flee the consequences of a drought seems to be false. Depending on their socio-economic position, they might have the choice between a variety of coping strategies, including migration. On the other hand, they might also be too poor to migrate at all, meaning that migration as a way of escaping their situation is as unlikely as any other strategy.

3 Methodology – A Qualitative and Holistic Approach

An investigation into the relationship between the variability in precipitation patterns and migration needs to cope with the methodological problem of isolating climate as one factor among many that influence migratory behaviour. Using a qualitative and holistic approach is one way to overcome this issue by seeking to gain understanding from the perspective of the people concerned. The aim of my study is, therefore, to understand how different groups of people perceive the stress caused by changes in the annual rainfall pattern and how they cope with it. A second group of questions is how different groups of people make migration decisions, who decides, who migrates and who stays, where do people go and for how long and why. The theoretical background of the research builds on insights from the Sustainable Livelihoods Approach (SLA) and the New Economics of Labour Migration (NELM).[5]

Figure 8.1 demonstrates how different groups of people might cope in different ways with the consequences of changing rainfall patterns. Migration might be one of these strategies, but it is not the only one.

Figure 8.2 demonstrates different motives to migrate or not to migrate. Many migration decisions are likely to combine two or more of these motives. Some of them are sensitive to climate change, others are not. So, reasons why people migrate can be tested for their sensitivity to climate change. Apart from the question whether somebody wants to migrate or not s/he has to decide where to go, how long to stay and if s/he wants to go alone or with family members or friends.

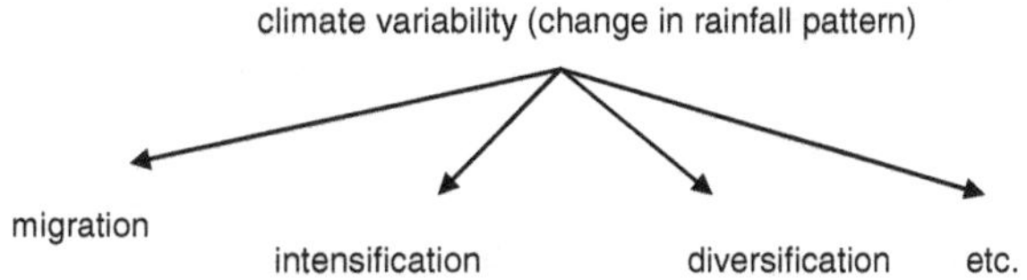

Fig. 8.1 Different coping strategies (source: author)

[5] A more detailed description of these concepts and how they link to my research can be found in Kniveton et al. 2008.

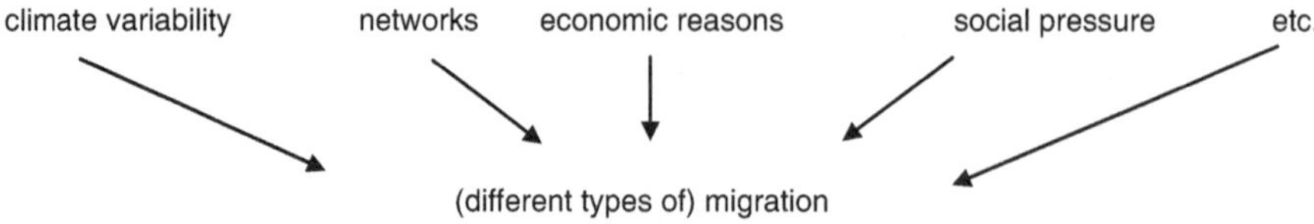

Fig. 8.2 Different motives for migration (source: author)

Looking at the relationship between climate variability and migration from these two angles ensures a holistic understanding of the situation. To this end, I conducted about 100 informal interviews in two communities with people from different socioeconomic levels, migrants and non-migrants, women and men and old and young people. The information gained in these interviews was completed by life histories, participant observation, expert interviews (doctors and teachers working in the communities as well as local politicians and scientists) and by local statistics.

4 The Research Setting – Migration and Climate in Zacatecas

The state of Zacatecas is situated in the Centre-North region of Mexico and labelled as a classical migration state of the country. The first migrants were contracted by the USA to build railways and to work in agriculture and mining more than a century ago. During the Bracero Programme from 1942 to 1964, more than 4.5 million Mexicans worked in the USA. Most of them originated from the four states of Jalisco, Michoacán, Guanajuato and Zacatecas (Massey et al., 1987: 39–44). Nowadays, Zacatecas is still one of the most important migrant-sending states in Mexico. According to the 2000 census of the Mexican Institute for Statistics, Geography and Informatics (INEGI), in 2000, 4.9% of the population in Zacatecas migrated to the USA, which is the highest rate of all states and far above the Mexican average rate of 1.6%.[6] Results of the 'Encuesta sobre Migración en la Frontera Norte de México' (EMIF), as shown in Table 8.1, indicate that after a decrease of migrants between 1995

Table 8.1 Flow of migrants from the state of Zacatecas into the USA by year

Year	1995	1999	2000	2001	2002	2003	2004	2005
Migrants	25,487	20,415	15,956	12,014	18,722	17,757	17,778	19,663

Source: Data elaborated by El Colegio de la Frontera Norte based on the longitudinal study 'Encuesta sobre Migración en la Frontera Norte de México'.[7]

[6] INEGI. XII Censo General de Población y Vivienda 2000. For further information see www.inegi.gob.mx. Data from the 2000 census is the most recent available information about migration from INEGI.

[7] For further information consult the website of the Colegio de la Frontera Norte www.colef.mx or the most recent publication 'Encuesta sobre Migración en la Frontera Norte de México' (2005).

and 2001, migration from Zacatecas to the USA increased again between 2002 and 2005, without reaching the numbers of 1995.

These numbers, as in most cases of migration statistics, should be treated carefully, because most of the border crossings are undocumented. Furthermore, the state of Zacatecas is not homogenous with respect to the intensity of international migration. As Delgado Wise et al. (2004: 168) point out, while 23 of the 57 municipalities in the state of Zacatecas show this very high international out-migration rate, 7 of the municipalities show a low or very low international migration rate (Fig. 8.3).

Zacatecas has always been coping with dry weather. For the period between 1941 and 2005, the Mexican National Meteorological Service (SMN) indicates an average annual amount of precipitation of 517.6 mm in comparison to 773.5 mm on the national average.[8] The majority of people asked in both of the researched communities in the state of Zacatecas say that over the last 15 or 20 years, the weather has become drier than it used to be. However, this perception is not confirmed by rainfall data on the state level. According to the INEGI (2007: 10), in 2006, total annual precipitation in Zacatecas was with 520.7 mm very close to the average observed by the SMN between 1941 and

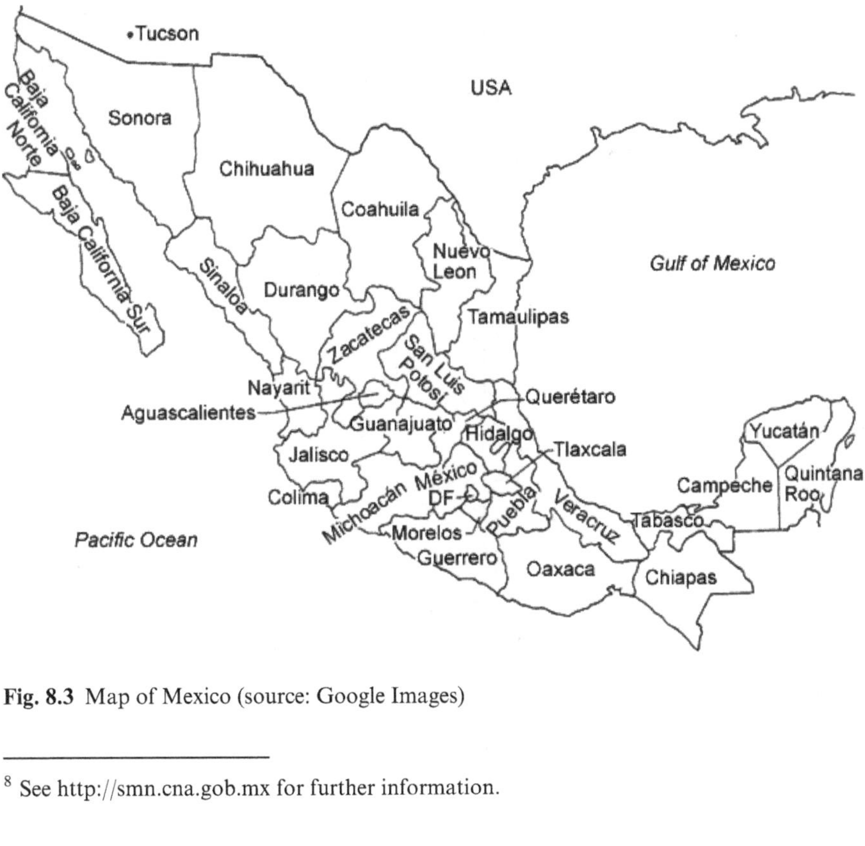

Fig. 8.3 Map of Mexico (source: Google Images)

---

[8] See http://smn.cna.gob.mx for further information.

Table 8.2 Precipitation in Zacatecas on a monthly basis (in mm)

	January	February	March	April	May	**June**	July	August	September	October	November	December
Average (1953–2006)	17.4	8.1	5.5	9.3	16.6	76.5	96.1	102.3	80.3	36.8	9.8	13.6
2006	24.0	0.0	0.0	1.5	12.4	38.5	125.3	104.3	114.4	66.0	11.5	22.8

Source: INEGI (2007: 10).

1996. It was even higher than the average between 1953 and 2006 of 472.3 mm of annual rainfall as observed by INEGI (2007: 10). Furthermore, people observe higher temperatures in the summer as well as a changing annual pattern of precipitation and temperatures. Rainfall usually occurred between June and August, and in the past, this rainfall pattern assured that farmers could cultivate their land in the summer and harvest before the cold period in November began. During recent years rain might start in August instead of June, which means that it is too late to assure a harvest before the frost, as the cold period in some years also starts earlier in the year than before. Furthermore, in recent years, rainfall – as well as hail that also destroys the harvest – is said to be heavier than in the past. People's perceptions of changing patterns of precipitation are confirmed by the precipitation statistics for Zacatecas as Table 8.2 shows, using the example of the year 2006 in comparison with the period between 1953 and 2006.

The summers of 2005, 2006 and 2007 have been exceptionally dry in Zacatecas and many farmers say that they have lost their harvest. In 2007, in addition, a period of cold days in the month of September affected agriculture and brought about a loss of the maize harvest. For 2008, the government of the state again anticipated a very dry summer and, therefore, many farmers did not cultivate their land. Because of the high prices for seeds, fertilisers and agricultural machinery, farming has become an investment that many people do not want to make anymore given the insecure rainfall conditions. However, those farmers, who did cultivate maize and beans in 2008, report a very good harvest compared to the previous years, and the amount of rainfall in 2008 was perceived to be higher than average.

5 Results

In order to understand how people cope with the changes they observed in the climate and the triggers of their migration decisions, I conducted a village-level study in two rural communities, Laguna Seca (municipality of Pánuco) and El Tigre – El Nuevo Tigre – Ojo de Agua del Tigre (municipality of Villanueva). These communities were selected, because they are distinct from each other regarding (1) the level of migration (Pánuco has got a very low US migration rate, Villanueva has got a very high US migration rate), (2) the socioeconomic level (many people in El Tigre receive remittances), employment opportunities (in El Tigre, there is no work at all, except for some jobs in the catering business in Villanueva. In Laguna Seca, many people work in commercial agriculture and some in a local brewery or in road construction) and (3) the level of dryness (El Tigre is very dry, but people can forage cactus and cactus fruit, and there are waterholes for the cattle. In Laguna Seca, the weather is even drier).

Analysing the behaviour of different groups of people in these two different communities showed the following results regarding people's livelihood strategies and migration decisions.

5.1 Livelihood Strategies

In Laguna Seca, the lack of rainfall is a factor that seriously affects subsistence agriculture. Other factors are rising prices of seeds and fertilisers as well as decreasing prices they obtain for their harvest. Combined with weather conditions that are getting more unpredictable, many people do not want to take the risk of a bad harvest and lose all their capital. This means that more families abstain from cultivating their fields. Therefore, they need to buy basic products such as maize and beans, which they used to cultivate in the past. As a consequence, they become even more dependent on the development of world market prices for food, which are also heavily affecting their livelihoods. Commercial agriculture is also affected by the changing rainfall and temperature patterns as well as the changing prices. Some landowners switched to cultivating pasture for the cattle, which hardly demands any labour, while others decided to cultivate more labour-intense products that are worth more on the market. The effects for the day workers are therefore not foreseeable yet. Many people think that life is economically more difficult nowadays than it was before. An important livelihood strategy is to 'live with what is available'. In the difficult months, with no work in the field, people do not buy clothes or household items but only the most essential food items. They also use farmer credits that some banks offer or credits provided by a few shops in the village. During that period, people who do not suffer from seasonal unemployment often give or lend money to family members who are unable to work or who need extra money in case of disease or accidents.

It is mostly women who invent strategies to generate income by selling homemade food, cheese, clothes, shoes, cosmetics, ceramic products, Tupperware, diapers, etc. Moreover, they offer services, such as making and repairing clothes, cutting hair and ironing. However, their income also depends on the work that is available in the village. In the times where there is no work in the fields there is less work for them due to the decreasing purchasing power. Furthermore, about 200 women form part of a government programme called 'Oportunidades', in which the participants receive 460 Pesos every 2 months and additional grants for the children who are attending school.

In El Tigre, remittances are certainly the most important source of income. They are – next to the ownership of cattle – a factor that distinguishes families by socioeconomic status. Families in which the father is sending remittances tend to have more money. Elderly couples receiving money from their sons or daughters who also have to maintain their own family are usually worse off. In this case as well, women are more active than men, as far as creating sources of income is concerned. Small shops offering basic food and drugstore items are owned and maintained by women. As many people own cows, many women are selling home-made cheese. Some are also selling food, others ceramic products. However, as the socioeconomic level here is higher than in Laguna Seca – mainly due to the remittances people receive – these activities contribute relatively less to the maintenance of the household. Some women who are better off, therefore, gave up these activities, because they do not feel that they pay off. Credits provided by banks are also used in El Tigre, and women participate in the 'Oportunidades' government programme as well.

5.2 Migration

Although Laguna Seca lies in an area with a very low international migration rate, people still migrate there. As is common in Mexican rural societies, women move to their husband's village after marriage. This means that in one community, there are a lot of women from neighbouring communities. Some men migrate internally during the times of non-employment in agriculture. However, they usually only stay for 2 or 3 months at most and return back. Most of these internal migrants go to the coast of Jalisco, a neighbouring state with tropical climate, or to the state of Sinaloa, also at the coast of the Pacific Ocean, to work in agriculture. A few of them also go to Zacatecas, Guadalajara or Monterey. There is some out-migration of young people to attend 'preparatoria' or University, either in Zacatecas or in Pozo de Gamboa, a neighbouring community. Some of them commute, others who have relatives in either place prefer to move there to avoid the daily bus fare. Some better-off families also decided to move to the city of Zacatecas to allow their children to study and live with their families. However, these cases are rare. There are some men travelling with work contracts to the USA, but even more to Canada. These contracts usually last for a few months, for which the contracted people obtain a visa; travel and housing expenses are covered by the employers. However, the costs for the passport and the visa have to be covered by the migrants, meaning that not many people can afford this form of migration. A few men go as 'mojados', which means that they cross the border illegally. In most cases, these people have some family connections to villages with higher international migration rates, for example, if their wives originate from such a place and they cross the border together with their brothers-in-law or other relatives of their wives.

There are many reasons why some people do not migrate; the most common answer is that they do not have any relatives, neither in Zacatecas, nor in other parts of Mexico or the USA. An illiteracy rate of about 50% among the adult population (according to the 2007 village census effected by the local health centre) is also a reason that people do not move easily. The lack of education causes the fear of not coping with the life in big cities, including Zacatecas. This is linked to the fear of losing what they have for some insecure outcome.

Another reason that often remains unspoken is that the socioeconomic conditions in the village suggest that most people do not possess the financial means for border crossing.

In El Tigre, as well as in Laguna Seca, married women move to their husband's villages, and some young people leave the village for school enrolment and higher education in Villanueva or Zacatecas. However, international migration is much more common in this community. Almost all migrants who leave in search for work go to the USA, the major destination being Chicago. People remember that the first migrants who left for the USA with work contracts did that 50 years ago. Today, almost every family within the villages has at least one member in the USA, if not many more. The link between unemployment and migration is obvious and mentioned by everybody in the communities. However, there are two general aspects to consider:

(a) There are different forms of migration. The most important distinction is between legal and illegal border crossings, the majority being illegal. Illegal border crossings are nowadays usually attempted by men. Most people say that it has become too dangerous for women and children. Women are still leaving the village, but very often their husbands arranged papers for them, after obtaining a USA residence permit. Linked to the illegal–legal distinction is the distinction between temporary and permanent moves. There are many people originating from the communities who live without papers in the USA and have not returned back for many years. The number of illegal migrants has decreased, since crossing the border has become more dangerous and expensive, and the security controls become stricter and more complicated.

(b) Not everybody is moving; people above 50 years and others suffering from diseases do not migrate anymore, because they believe that their chances in the US labour market are very low. Some other people are not keen on leaving for the USA, especially those who had bad experience in the past.

6 Conclusion

These preliminary results show that there is no obvious direct relationship between climate variability and migration; people are making use of a variety of livelihood strategies, and they have many reasons for migrating or for staying in their village. People perceive the climate as changing but they do not perceive migration as an option to cope with these changes. The most important reasons they indicate for their moves, whether internal in Laguna Seca or international in El Tigre, are unemployment and low wages for the few jobs that are available close to their homes.

The comparison between the two communities clearly shows that migration networks that developed as a consequence of the village history of migration are the most important factors that enable people to migrate internationally. The dry weather might have been a factor that contributed to provoke migration in El Tigre 50 years ago. Afterwards, migration became part of people's life, since for some people, it has become a necessity to survive, and for others, a way to maintain their standard of living. However, people think that crossing the border as undocumented migrants has become more difficult, dangerous and expensive in recent years and fewer people manage to cross the border. As a consequence, also fewer people return to El Tigre because they fear they would not be able to cross the border into the USA again if they want or need to do so later. Thus, while weather conditions are perceived to be more severe than in the past, fewer people are migrating internationally.

Laguna Seca does not have a history of international migration and, therefore, very few people migrate internationally. It is unlikely that people in Laguna Seca will manage to break into the system of international migration now that the border crossing has become more difficult. Village dwellers depend to a larger extent on farming than in El Tigre because of the employment opportunities in commercial agriculture and because hardly any family is receiving remittances. This means that they are suffering to a larger extent from the climate variability that they observed as affecting agriculture. Some farm workers in Laguna Seca have already migrated internally during the months in which no jobs in agriculture are available. It is conceivable that more people might be forced to do so and that they might be forced to do so for longer periods in the year under a scenario of worsening changes of the climate. However, as the majority of internal migrants in Laguna Seca are looking for jobs in agriculture, they might also be affected by changes of agricultural productivity caused by climate change in the receiving regions. Furthermore, as many villages are likely to suffer from the same problems, receiving areas might also get saturated.

It seems thus unlikely that international migration will increase as a consequence of future climate change while the effects of climate change on internal migration are not foreseeable yet.

Acknowledgements This chapter presents the preliminary results of an empirical study that I conducted from January to June 2008 in two rural communities in the Mexican state of Zacatecas. This research is part of my DPhil project at the University of Sussex, UK, entitled 'The potential influence of climate change on migratory behaviour – a study of drought, hurricanes and migration in Mexico'. The presented work has benefited a lot from the ideas of my thesis supervisors Prof. Richard Black and Dr. Dominic Kniveton. The sections of this article that describe the theoretical background and the methodology of the study have been published in a similar form before:

Kniveton, D., K. Schmidt-Verkerk, C. Smith, and R. Black. (2008): *Climate Change and Migration: Improving Methodologies to Estimate Flows*, IOM Migration Research Series Paper No. 33, 2008, International Organisation for Migration, Geneva.

The description of the study does also partially overlap with another earlier publication:

Schmidt-Verkerk, K. (2009): The potential influence of changing precipitation and temperature patterns on migratory behaviour in the state of Zacatecas, Mexico. In: Oliver-Smith, A., and Shen, X. (eds.), *SOURCE No. 12/2009: Linking Environmental Change, Migration, and Social Vulnerability*, United Nations University Institute for Environment and Human Security, Bonn.

References

Black, R. (2001): Environmental refugees: Myth or reality? UNHCR *Working Paper No. 34*, Geneva.

Black, R. (1998): *Refugees, Environment and Development*. Addison Wesley Longman Limited, Harlow.

Boyle, P., K. Halfracree, and V. Robinson. (1998): *Exploring Contemporary Migration*. Pearson Education Limited, Harlow.

Castles, S. (2000): *Migration as a Factor in Social Transformation in East Asia*, Paper presented to a Conference on Migration and Development, Princeton University, 4–6 May, 2000.

Castles, S., and M.J. Miller. (1993): *The Age of Migration: International Population Movements in the Modern World*. Macmillan, Basingstoke.

Christian Aid. (2007): *Human Tide: The Real Migration Crisis – A Christian Aid Report*. Christian Aid, London.

De Haan, A. (2000): Migrants, livelihoods, and rights: The relevance of migration in development policies. Social Development *Working Paper No. 4*. Department for International Development, London.

Delgado Wise, R., H. Márquez Covarrubias, and H. Rodrígez Ramírez. (2004): Organizaciones Transnacionales de Migrantes y Desarrollo Regional en Zacatecas. *Migraciones Internacionales* 02(2/004): 159–181, Colegio de la Frontera Norte, Tijuana.

El-Hinnawi, E. (1985): *Environmental Refugees*. United Nations Environment Programme, Nairobi.

Encuesta sobre Migración en la Frontera Norte de México. 2005, 2007 edition, Mexico.

Ezra, M., and G.E. Kiros. (2001): Rural out-migration in the drought prone areas of Ethiopia: A multilevel analysis. *International Migration Review* 35(3), 749–771.

Findley, S.E. (1994): Does drought increase migration? A study of migration from rural Mali during the 1983–1985 drought. *International Migration Review* 28(3), 539–553.

Google Images. (2007): Available from http://images.google.co.uk/images?q = map + of + mexico&hl + en&rls + GGLJ,GGLJ:2006-35,GGLJ:en&um = 1&sa = X&oi& = images &ct = title, visited on 30 August, 2007.

Haug, R. (2002): Forced migration, processes of return and livelihood construction among pastoralists in northern Sudan. *Disasters* 26(1), 70–84.

Henry, S., B. Schoumaker, and C. Beauchemin. (2004): The impact of rainfall on the first out-migration: A multi-level event-history analysis in Burkina Faso. *Population and Environment* 25(5), 423–460.

INEGI. (2007): *Anuario Estadístico Zacatecas, Edición 2007*. Instituto Nacional de Estadística, Geografía e Informática, Aguascalientes, Mexico.

IPCC. 2007: Climate Change 2007: Synthesis Report. Contribution of Working Groups I, II and III to the Fourth Assessment Report of the Intergovernmental Panel on Climate Change. Core Writing Team, Pachauri, R.K., and Reisinger, A. (eds.), IPCC, Geneva.

Jakobeit, C., and C. Methmann. (2007): *Klimaflüchtlinge – Die verleugnete Katastrophe*. Greenpeace, Hamburg.

Kibreab, G. (1997): Environmental causes and impact of refugee movements: A critique of the current debate. *Disasters* 21(1), 20–38.

Kniveton, D., K. Schmidt-Verkerk, C. Smith, and R. Black (2008): Climate change and migration: Improving methodologies to estimate flows. *IOM Migration Research Series Paper No. 33*, International Organisation for Migration, Geneva.

Kritz, M.M., Lin, L.L., andZlotnik, H. (eds.) (1992): *International Migration Systems: A Global Approach*. Clarendon, Oxford.

Lonergan, S., and A. Swain (1998): *The Role of Environmental Degradation in Population Displacement*, Environmental Change and Security Project Report, Issue 4, GECHS, Washington, DC.

Massey, D., R. Alarcón, J. Durand, and H. González. (1987): *Return to Aztlan: The Social Process of International Migration from Western Mexico*. University of California Press, Berkeley, CA and Los Angeles, CA.

McGregor, J. (1993): Refugees and the environment. In: Black R., and Robinson V. (eds.), *Geography and Refugees: Patterns and Processes of Change*, Belhaven Press, London.

Meze-Hausken, E. (2004): Migration caused by climate change: How vulnerable are people in dryland areas? *Mitigation and Adaptation Strategies for Global Change* 5(4), 379–406.

Myers, N., and J. Kent. (1995): *Environmental Exodus, an Emergent Crisis in the Global Arena*. Climate Institute, Washington, DC.

Saunders, P.L. (2000): Environmental refugees: The origins of a construct. In: Stott P., and Sullivan S. (eds.), *Political Ecology: Science, Myth and Power*. Arnold, London.

Skeldon, R. (2002): Migration and poverty. *Asia-Pacific Population Journal* 17(4), 67–82.

Stern, N. (2006): *The Economics of Climate Change – The Stern Review*. Cambridge University Press, Cambridge, MA.

Suhrke, A. (1994): Environmental degradation and population flows. *Journal of International Affairs* 47, 473–496.

UNFCCC (1992): United Nations Framework Convention on Climate Change, United Nations.

Wood, W. (2001): Ecomigration: Linkages between environmental change and migration. In: Zolberg, A., and Benda, P. (eds.), *Global Migrants, Global Refugees: Problems and Solutions*. Berghahn Books, New York, NY.

Part IV
Migration and Other Forms of Environmental Degradation

Since the United Nations Conference on the Human Environment in 1972, which led to the establishment of the United Nations Environment Programme (UNEP), environmental issues have been increasingly visible on the international political agenda. Twenty years later at the Earth Summit of Rio de Janeiro in 1992, agreement was reached on the United Nations Framework Convention on Climate Change (UNFCCC) and the Convention on Biological Diversity, both of which were subsequently ratified and became legally binding. Despite these and a wide range of other international agreements on environmental issues, including the protection of the stratospheric ozone layer, combating desertification, and reducing the long-distance transport of atmospheric pollutants, the Earth is still facing a large number of complex environmental challenges.

The possible connection between migration and environmental changes has also been discussed since the 1970s, and several chapters in this book refer to some of the relevant literature. While environmental experts framed the issue in terms of the dire consequences of environmental change and possibly large numbers of migrants, migration experts focussed on the economic pull and push factors that trigger migration and rarely considered that environment could be a factor.

More recently, experts from natural science and social science disciplines have begun to work together to study the possible linkages between environment and migration (see, in particular, Part V of this book). This research is showing the complex interlinkages between the various factors that lead to decisions to migrate. Much academic and media attention has been given to the possible role of climate change in causing migration (see also Part III of this book), but in this part we focus on other types of environmental change.

The first chapter deals with rain-fed regions in India, where vulnerable populations who rely on farming activities are forced to migrate due to land degradation and sometimes even suffer from rural violence, despite the enforcement of the Watershed Based Development Program which was originally designed to ensure and enhance rural livelihoods, management of natural resources, and conservation of soil and water. The second chapter takes the regions Durres in Albania, Rudňany in Slovakia, and Mitrovica in Kosovo, as three case studies where migrants and refugees from other regions/countries are exposed unequally to the polluted environment compared to the nationals and local communities. The chapter also highlights the paradoxical situation in which people actually move into seriously polluted areas as well as competition and conflict between two groups of unequal social and political status over access to a safe and healthy environment. The third chapter enquires whether relocation due to natural hazards would reduce vulnerability or cause more problems to the relocated persons. Focusing on the tsunami of December 2004 and the consequent relocations in Sri Lanka, the chapter calls for careful planning before displacing the affected populations, in order for the latter to have appropriate access to job and education opportunities as well as social networks in the new settlements. The final chapter of this part gives a review of historic and recent relocation processes in Guatemala that were caused by natural disasters. It also provides policy-oriented recommendations for successful relocations.

Chapter 9
Environmental Migration from Rainfed Regions in India Forced by Poor Returns from Watershed Development Projects

Kaushalya Ramachandran and Padmaja Susarla

1 Introduction

In India, watershed-based development has been the strategy for growth and sustainability of agriculture in semi-arid and dry sub-humid regions popularly termed rainfed agro-ecological regions (Fig. 9.1) (Velayutham et al. 1999). The strategy has been developed to ensure and enhance rural livelihoods, management of natural resources and conservation of soil and water that attracted large public investments in the last 30 years. More investments have been assigned for the same purpose in recent times to ensure sustainable development in rainfed areas. Rainfed agro-ecological regions extend over 76.74 million ha out of the total geographical area of 90.4 million ha in the states of Andhra Pradesh, Maharashtra, Karnataka and Tamilnadu in Peninsular India. According to the Census of India – 2001, over 71 million persons are involved in agriculture and related activities in these four States alone, which contribute a large amount of food grain, pulses and oilseeds to the national food stock. As perennial irrigation sources are lacking and mean annual rainfall ranges from 300 to 800 mm (occurring on 52–55 rainy days) and is also associated with thunderstorms that induce massive soil erosion and land degradation, a watershed-based development programme has been adopted as the major development initiative for improving rural livelihoods in the region (Katyal et al. 1996). A relentless increase in population and the national priority of achieving food grain self-sufficiency proscribes a growth rate of 4.0–4.5% annually (Planning Commission, 2007) compared with the current rate of growth of 2.72% (MoF, 2008). This increase seems insurmountable in view of the falling public investment coupled with cost escalations, unhindered land degradation and competing demand for land for other uses. Even popular political gimmicks, such as subsidised or free electric power for the farming sector and subsidies for other agricultural inputs are failing to bind the farmers to their land, resulting in an exodus from rural areas to neighbouring urban areas.

K. Ramachandran (✉)
Central Research Institute for Dryland Agriculture, Hyderabad, India
e-mail: kausalyar@yahoo.com

T. Afifi, J. Jäger (eds.), *Environment, Forced Migration and Social Vulnerability*,
DOI 10.1007/978-3-642-12416-7_9, © Springer-Verlag Berlin Heidelberg 2010

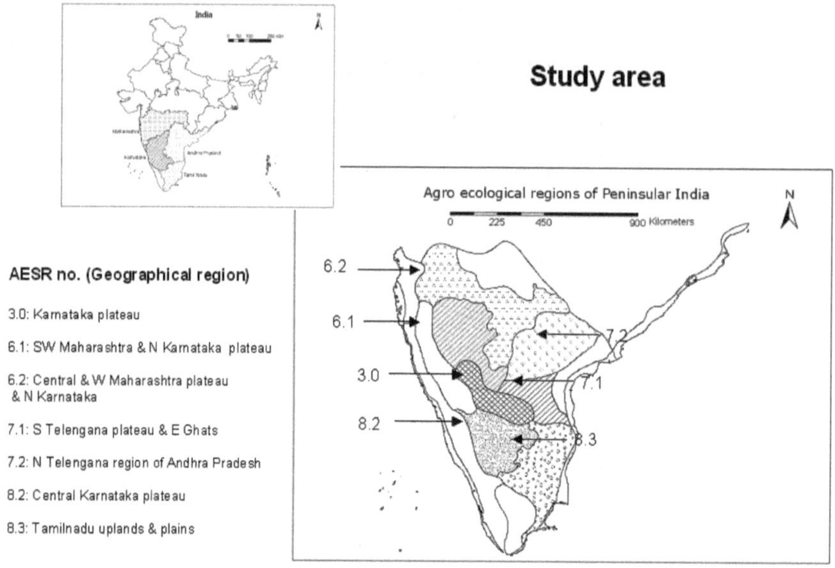

Fig. 9.1 Extent of rainfed agri-ecosystem in peninsular India (source: Velayutham et al., 1999)

This is shown by the data on migration of rural people in the region. In Andhra Pradesh migration has increased since 1971. This is evident within and from outside the state. There has been a marked increase in the rate of migration since 2001. In 1991, 20% of the rural population in the state migrated, and by 2001 this figure increased by 1%, with the actual number of persons who migrated being over 16.02 million. On the other hand, urban migration was less than half of this figure, approximately 7.4 million persons. Understandably, this migration is due to the deterioration of agriculture and loss of sources of livelihood in the rural areas, forcing migration of rural people. Another peculiar aspect of this is the higher level of female migration,contrary to the trends seen in the states of Maharashtra and Orissa, where young males migrate to urban areas in search of livelihoods, leaving behind the elderly and women. The analysis of reasons for migration of rural females reveals that matrimony was a major reason for migration, because eligible males in rural areas have poorer livelihood options, thus making alliance with them unattractive. In the case of urban migration, employment, marriage and family matters were equally important factors influencing the decision to leave. Thus, environmental factors that adversely affect the remunerative capability of agriculture seem to be more responsible for migration in the rainfed regions in Peninsular India (Ramachandran and Padmaja, 2005).

Since the 1980s, development projects like the Watershed Development Programme have been implemented to conserve and manage land resources in the rainfed region at a massive cost. However, these projects seemingly failed to ensure remunerative agriculture and provide a good livelihood to rural people, making them vulnerable to climatic vagaries and forcing them to migrate to urban centres with poor coping skills to live and work in an alien and often hostile environment there. These areas are ill-equipped to absorb the increased human load. As land within urban areas is costly, the migrants are forced to live in slums set up on available open spaces within the urban areas or in the urban–rural divide zone that turns into a ground of strife.

This chapter aims to present the unique situation of agricultural production systems in the region, so as to highlight the causes of migration. The climatic variability and land degradation have been described as drivers of environmental migration and the social vulnerability of rural populace of the rainfed region has also been described. The chapter clearly indicates how climatic variability makes agriculture vulnerable in a region perennially threatened by climatic aberrations. The study highlights the necessity of implementing a watershed development programme in rainfed areas to backstop the agricultural sector.

1.1 Watershed Development Programme in India

The southwest monsoon is the primary rainfall season and source of water for the rainfed agro-ecological region of Peninsular India. However, most of the rainfall, i.e. 450–800 mm, occurs in 50–55 days, although the number of days has been as low as 31 in 2004 which was a major drought year and 88 days in 1983 a normal year (Katyal et al. 1993). Most of the rainfall events are also associated with thunderstorms that last for less than an hour, leading to soil erosion, flooding and overland runoff, and causing in turn severe loss of topsoil and soil nutrients. In view of such a rainfall pattern, the watershed-based development project was considered crucial for water harvesting, storing and use for critical crop phenophases in agriculture, besides undertaking other related activities like fishing, livestock rearing, horticulture, market gardening, etc., for rest of the year. The Watershed Development and Management Programme was initiated in India in 1973–1974 by the Department of Land Resources, the Ministry of Rural Development (MoRD), the Government of India under the banner of Drought Prone Area Programme (DPAP). The Ministry of Agriculture (MoA) undertook a watershed development programme in 1974–1975 to check soil erosion in areas where shifting cultivation was rampant (Planning Commission, 2007). Since 1988–1989, the MoRD initiated the Integrated Watershed Development Programme (IWDP) while the National Watershed Development Programme for Rainfed Area (NWDPRA) was initiated by the MoA in 1990–1991. The key aspects of the watershed development programme (WDP) were the development of Soil and Water Conservation (S&WC) structures of check-dam, stone weirs, contour bund, live bunds, vegetative cover, key-line plantation, grass way, etc., to provide impediments to overland runoff that induces soil erosion and depletes nutrients from agricultural fields. Structures were laid to guide runoff to designated farm ponds and tanks for water harvesting on the surface, besides impounding water for facilitating deep percolation for groundwater recharge. Thus, WDP was considered to be the most comprehensive programme for achieving agricultural and ecological sustainability in the rainfed regions of India. A hallmark of the watershed development programme was the implementation of improved land management practices for each aspect of agriculture and rural life in rainfed regions. Through watershed projects till 2007, a total of 50.8 million ha is considered as treated in the country at a total cost of USD 4,278 million (Planning Commission, 2007).

As India aims to achieve a sustainable agricultural growth rate of 4.0–4.5% within the current decade, in order to reduce food insecurity and poverty while increasing rural purchasing power, an evaluation of the performance of WDP indicated some disquieting trends. The present chapter describes these trends indicating how and why the watershed projects fell short of the expectations and the consequences of this failure vis-à-vis agricultural production and income security in the rural areas.

1.1.1 Evaluation of Watershed Development Programme

The WDP has been evaluated by numerous studies, such as Kerr et al. (2002), Shah et al. (2004), Joshi et al. (2005) and Sreedevi et al. (2004). In addition to these studies, there are a few policy review studies on the watershed development of the Andhra Pradesh Government by Oliver Springate-Baginski et al. (2004), of the Watershed Development Programme in India by Hanumantha Rao (2000), and of the Sustaining Rural Livelihood in Fragile Environments by Ratna Reddy et al. (2004). Most of these studies have evaluated the impact of a package of practices implemented under WDP and were based on qualitative data with some quantitative information for which econometric analysis had been performed (Kerr et al., 2002). Joshi et al. (2005) undertook a meta-analysis of over 311 watershed projects implemented in India and documented their efficiency, equity and sustainability benefits. The authors pointed out that the mean benefit to cost ratio of a watershed programme in the country was quite modest at 2.14. The internal rate of return was 22% that was comparable with any rural developmental programme. The study concluded that rural development projects like the watershed development project could not be expected to yield better results. This cynicism has adversely affected the future prospects of the agricultural sector in India and rainfed agriculture in particular.

In order to assess the impact of watershed development projects and to identify the reasons for their limited success, a research programme was initiated in 2005 with the funding of the Indian Council of Agricultural Research (ICAR) to study the impact of watershed development projects in 10 micro-watersheds in the Northern Telangana agro-ecological region. This region is a typical rainfed area where agriculture is the main economic activity in the rural region. The study was strengthened with tools of Geomatics and supplemented with conventional methods of study like soil fertility assessment, socio-economic survey and econometric analysis. Over 50 sustainability indicators were constructed and the evaluation studies have brought out some disconcerting facts (Ramachandran et al. 2007).

Despite the heavy investment to improve agricultural productivity, crop yields have not risen sharply. In the study of watersheds paddy yield ranged from 3 to 5 t ha^{-1} in treated micro-watersheds, while the district average for *Rangareddy* was 2.59 and 3.1 t ha^{-1} in the case of the *Nalgonda* district. Yields in untreated micro-watersheds, where traditional practices were followed, were found to be 2.5–4.75 t ha^{-1} (DES, 2006; DoAC, 2004). This is lower than the levels achieved in Tamilnadu and Karnataka and far lower than the yield levels in other parts of the world. Although sorghum yield was double that of the District Average, its usage is falling due to distribution of rice through the Public Distribution System in the country. Similarly, the yields of cotton, oilseeds and pulses were lower. This resulted in the shifting of prime agricultural land to other uses and to real estate development in the vicinity of the Hyderabad Urban Agglomeration (HUA) (Ramachandran and Padmaja, 2005).

Several progressive farmers have begun to cultivate vegetables and flowers for markets of the HUA region. This resulted in increased income for these farmers with the average income of small farmers with 1–2 ha of land holding ranging from $ 54 to 517 during *Kharif* or the southwest monsoon season in 2006. This is lower than the income earned by migrant labourers in the urban centres in the vicinity. It has also caused an exodus of marginal and small farmers in search of odd jobs like sweeping roads, undertaking small masonry work or joining the large tribe of casual workers at large construction sites. These estimations were made based on an income costing method called *Cost A1* where imputed values of family labour, rent for leased land or interest for capital, etc., are not considered. When these were also considered as in the case of *Cost B* and *Cost C*, the income was found to be negligible, making rainfed agriculture an unattractive option for livelihood (DoAC, 2003).

The average income $person^{-1}$ day^{-1} based on income computed from *Cost A1* was calculated for farmers in each micro-watershed and it was found that in treated micro-watersheds income ranged from 10 to 25 cents/day. It was lower in case of semi-medium and medium farmers owning 2–4 and 4–10 ha of rainfed agricultural land, respectively. Undertaking agricultural operations was costly and as a result the *B to C ratio* ranged from 0.19 to 5.16 in the treated watersheds. The average *rate-of-turnover* that indicates the ratio of gross income over total assets was also calculated. The results indicated that agricultural sustainability was a distant hope and that marginal farmers with poor assets could gain higher income through other labour, while all other categories of farmers who had more than 2 ha of land holding had lower *rate-of-turnover*. Equity was also found to be low in all the treated watersheds, thus indicating that the watershed development programme had limited impact on agricultural sustainability. The present trend indicates unsustainable use of natural resources for agriculture in the study area.

In view of this situation, it was deemed prudent to analyse the trends in migration of rural people, the drivers of environmental migration and the social vulnerability and political causes for migration.

2 Migration Process in Rainfed Regions in India

Migration data are collected in India along with the Population Census for each decade. Due to the sheer size of this exercise in a democratic country like India, the provisional data are released within a year of the Census followed by compilation and publication of data on various aspects of demography. Usually, the migration data are compiled and published after the census operation is over and usually takes 4–5 years until it is available for the public. This time lag slightly degrades the utility of the data, as quick responses to varying trends are not possible. The analysis of migration data of Andhra Pradesh, a major state in peninsular India with Hyderabad being the fifth largest metropolis in India, indicates some interesting trends; the population structure of the various districts around the Hyderabad metropolis indicated that while Hyderabad had a working population, with 1.03 million main workers, over 95% of them worked in the urban tertiary sector. In the adjoining Rangareddy district, this group constituted over 58% or 580 thousand persons, but in neighbouring districts of Mahabubnagar, Warangal, Medak and Nalgonda, the rural work force belonging to main cultivator and main agricultural labour classes predominated (Fig. 9.2). In the case of marginal workers, it was seen that in the Hyderabad metropolitan region over 88% of them were engaged in the tertiary sector, while in the Rangareddy district, this group constituted the second largest after the dominance of marginal agricultural labour. In all other surrounding districts, marginal agricultural labour constituted 69–75%.

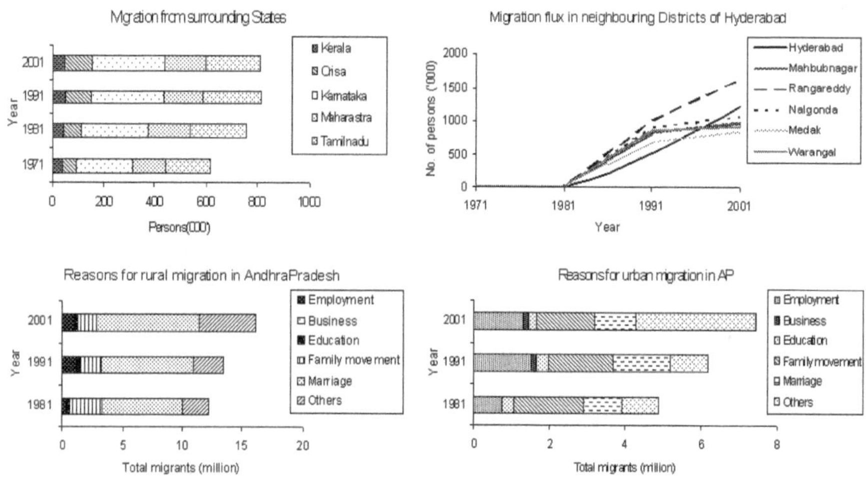

Fig. 9.2 Migratory trends in Andhra Pradesh (source: Census of India, 2001 – Andhra Pradesh)

While analysing the migration pattern of the literate population in Andhra Pradesh, it was seen that of a total of 39.2 million migrants, the share of illiterate migrants was 31%, while marginal workers, main workers and literates accounted for 25, 23 and 19%, respectively. In rural areas, illiterates and main workers were a large constituent, while in urban areas; literates and non-workers formed the larger segment of migrants.

Among literate migrants amounting to 5.33 million in rural areas, 52% were males and the rest females while in urban areas, this percent was more evenly divided among the 2.07 million migrants. Among the recent migrants (1–4 years), 55% of male literates and 45% of female literates moved. In rural areas, the trend was similar, while in urban areas, female migrants were almost equal to their male counterparts, i.e. 49%.

Among the main workers in Andhra Pradesh, 42% of the migrants were males and 58% were females. However, in rural areas, migration by main workers accounted for only 38% of the total of 7.99 million, while in urban areas, 68% of the total number of males (1.13 million) had migrated. Among marginal workers, female migrants accounted for 98% of all migrants. Among recent migrants (1–4 years) in the category of main workers in Andhra Pradesh, 83% were males and 17% females. In rural areas, the male ratio was 82%, while it was 85% in urban areas. Recent migration among male marginal workers was 17% in the state, although this ratio was 23% among urban men. However, female marginal workers were the dominant migrant population, according to Census (2001).

Analysis of intra-state migration in peninsular India indicated that the maximum migration into Andhra Pradesh was from the states of Karnataka and Tamilnadu, followed by Maharashtra. Migratory flux was felt most in the district of Rangareddy in 2001, followed by Hyderabad, Warangal, Mahabubnagar and Nalgonda.

The analysis of reasons mentioned for migration indicated that among females matrimony was the single most important factor for migration during the decades of 1980, 1990 and 2000. Among males, movement by families, employment and other reasons were mentioned as reasons for migration. The reasons (employment, education, business, movement of family, etc.) for rural migration, which has grown at a fast rate till 2001, have decreased in importance, while marriage and other reasons have gained importance for the migration decision. In the case of urban areas, employment, family movement, matrimony and other social and economic reasons influenced migration.

By 2001, over 12.0 million literates and 11.44 million illiterates in Andhra Pradesh moved, 9.8 million of who were main workers, 2.4 million marginal workers and 11.24 million non-workers. In the case of illiterates' migration, main and marginal workers were found to be more in rural areas, while in urban areas literates and non-workers formed a huge segment.

At a village level, it was seen that female migrants belonged to the 15–40 years age group, while male migrants moved till the age of 50. There were more male migrants in areas where no watershed development projects were implemented. Among the types of migration, seasonal migration was more as compared to permanent migration and more obvious in the case of untreated micro-watersheds.

3 Causes of Migration

There are several causes for migration in the rainfed regions of India and many of them are related to the prevalent agricultural production systems and the poor natural resource base in the region. As a result, low yield levels and consequently poor economic returns force migration. For instance, while average paddy yield among marginal farmers owning 1 ha of land is lower in the rainfed region, i.e. 2.3 t ha^{-1} compared to the state mean of 3.14 t ha^{-1} in 2006, its yield level improved among semi-medium and medium level farmers with 2–4 and 4–10 ha of land for cultivation. Yield levels were lower even in the case of red gram, sorghum and particularly maize, although it was slightly better in the case of cotton.

Consequently, returns from agriculture and related activities were lower. For instance, among marginal farmers, average annual income was $335 that prominently constituted wages from labour, followed by income from livestock. Among small farmers, the income fell marginally to $330. In the case of semi-medium and medium farmers where income from agriculture and livestock become significant, it rose to $358 and $560, respectively. The cost of inputs was found to be on an average $67 in the case of marginal farmers, increasing to $130 and $164 in the case of small and semi-medium farmers and lower, i.e. $ 142 in the case of medium farmers. This trend is due to the ownership of the resources at the disposal of the various categories of farmers that does not include the imputed values of land and labour. Marginal farmers practice subsistence farming and hence their costs are lower. The failure of dug wells and increased use of bore wells encouraged by free electricity supply to the farming sector by the State Government of Andhra Pradesh have resulted in a drastic fall in the ground water table. As a result, on average, water is extracted from a bore well for 8–10 h a day during the southwest monsoon season also called the *Kharif* cropping season, occurring during the months of July–September. Consumption is mostly in the Rangareddy and Medak districts. In addition, the study has indicated that the extent of surface water spread has also drastically fallen in and around the Hyderabad region. For instance, in 1971, the total number of water bodies in an area of 3,000 km^2 around the Hyderabad urban agglomeration was 1,271 in number. However, by 2000, this number fell to 960 (Ramachandran, 2001). Such a reduction in surface water storage has negatively affected ground water recharge and resulted in a failure of shallow bore wells. This has forced marginal and small farmers to abandon agriculture and migrate to urban centres in search of a livelihood.

4 Climatic Variability as a Driver of Migration

Peninsular India essentially has a semi-arid to dry sub-humid climate with a mean annual temperature ranging from 25 to 29°C and mean rainfall of 740 mm. The analysis of temperature data from 1951 till 2007 indicates that there is a rise in the mean maximum temperature from 32.2 to 32.5 and mean minimum temperature from 20.5 to 20.9 (Fig. 9.3). Due to high temperatures, evaporative losses are high and irrigation requirements increase when availability of water is a constraint.

The rise in temperature has also affected rainfall patterns in the region. The study indicates that since 1950 there have been 14 instances of El Nino in the Telangana region that forms the northern part of the state of Andhra Pradesh where Hyderabad is located; the latest event occurred in 2005. During the same timeframe, there were 12 events of La Nina, the last being in 2001. On more occasions, El Nino has caused a decrease in rainfall. Thus, while the mean monthly rainfall was 20 mm month^{-1}, in 2005, an El Nino year, there was a deficit in the mean monthly rainfall, i.e. 16 mm month^{-1}. In fact, in 2005, the other two regions of Andhra Pradesh, namely, Coastal Andhra and Rayalaseema also suffered from a deficit in rainfall. In Coastal Andhra, which is the rice bowl of Andhra Pradesh, a rainfall deficit of −25.71 mm/month in 2005 resulted in severe hardship for farmers. However, in the Rayalaseema region that is the drier part of the state, deficit rainfall in 2005 amounted to −88.5 mm/month (Vittalmurthy et al. 2008). Long-term analysis of the monthly trend

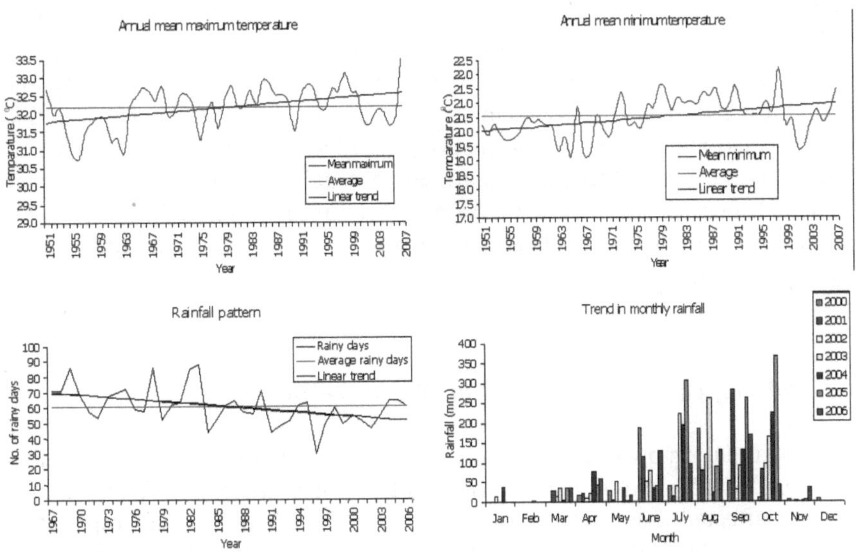

Fig. 9.3 Temperature and rainfall variability in Hyderabad region (source: CRIDA Annual Report 2006 and 2007)

indicates a slight positive rise in the case of Coastal Andhra and Telangana and a deficit in the case of Rayalaseema.

Analysis of inter-annual rainfall pattern indicated that there was a slight decrease in the mean rainfall from 755 to 742 mm (Ramachandran et al. 2007). Study of intra-annual rainfall patterns indicated that there was a shift in onset and withdrawal of the monsoon in the region. While the onset has shifted from June 12 to the last week of June, the withdrawal has moved from the end of September to October. Since 2000, maximum rainfall has shifted from July to August (CRIDA, 2006, 2007). These variations have adversely affected the agricultural calendar and created hardship, not only for farmers but also for the agricultural research community, as new seed material has to be developed urgently to suit the shorter growing period and higher temperature regime.

In order to assess the impact of watershed development projects in the context of rainfall variations, a study of the vegetative index was undertaken. Data from the Indian Remote Sensing Satellites (IRS) 1C & 1D LISS 3 were procured for two seasons; pre- and post-monsoon from 1998 till 2006, and the Normalised Differential Vegetative Index (NDVI) was analysed. It was found that the watershed projects that had been implemented at a huge cost had provided limited benefit in the form of vegetative cover, which is vital for fodder, incorporation in soils and protection from desiccation and erosion as the number of rainy days had decreased. This augurs ill for agriculture, as the intensity of rainfall has increased, which could accentuate soil erosion on the one hand and cause flooding on the other. The watershed projects were implemented precisely for managing such a situation by checking soil erosion, developing soil and water conservation structures and harvesting of water in surface structures like farm ponds and dugout wells for irrigation and deep percolation for ground water recharge. Due to election promises like free electricity supply for the farming sector from 2004, the water table has fallen drastically and agriculture had to be abandoned in several areas. In 2006–2007, it was seen that in the treated micro-watershed measuring 91.4 ha in Gollapalli, over 24 ha lay vacant, as the landholders had migrated to a neighbouring township. In an untreated micro-watershed in the same village measuring 88.7 ha, 22 ha lay vacant for the same reason. This trend was more or less similar in other watersheds in the region.

5 Land Degradation as a Driver of Migration

Land degradation owing to water and wind erosion is extensive in India. Water erosion affects 57.16 million ha in the country, while waterlogged areas, ravines and salt-affected areas extend to 3.20, 2.68 and 6.32 million ha, respectively (Singh et al., 2000). Wind erosion affects 10.46 million ha in the vast deserts of Rajasthan. Even the forest cover is degraded and covers 24.90 million ha. The situation in the rainfed region is worse. In Andhra Pradesh, out of the total geographical area of 27.44 million ha, over 12.23 million ha are degraded and face problems of water-induced erosion. Soil erosion due to running water is rampant because the topsoil, which is critical for agriculture, is less than 15 cm and in most places it is gravely to stony. Thus, in most of the land parcels where rainfed agriculture is practiced, the land is generally not usable for intensive agriculture.

The soil fertility status is poor, and our studies indicated that most soils in treated and untreated watersheds had a nitrogen content of <280 kg ha^{-1}. Similarly, other major soil nutrients like potassium and phosphorous and soil organic carbon content are low, which requires farmers to use fertilisers and manure. In view of the poor economic status of marginal and small farmers, the level of soil nutrients remains low and as a consequence yield and income also remain low, as indicated earlier in this chapter. All of these natural and environmental constraints make agriculture difficult and force farmers to migrate in search of a livelihood.

All major rivers of peninsular India, namely, Cauvery, Krishna and Godavari are rain dependent and hence, watershed projects are particularly critical for the undulating plateau region. Even 600 years ago, during the Vijayanagar empire period, much emphasis was laid on developing surface water storage structures, such as tanks, ponds and a vast network of canals. However, nowadays, many of these have crumbled owing to lack of maintenance and short-sightedness of the governing machinery. The two reservoirs built over a century ago by the Nizam, southwest of Hyderabad, namely, Himayat Sagar and Osman Sagar are also shrinking. Studies show how the extent of Himayat Sagar has decreased from 1,984.7 ha in 1971 to 827.3 ha in 2001 (Ramachandran and Padmaja, 2005). In 2004, when a severe drought was experienced, the extent of water spread in the reservoir was only 180.9 ha. Such drastic falls in surface water area affect the water table and cause immense hardship to people who depend on agriculture and livestock rearing for a livelihood and are forced to migrate.

6 Social Vulnerability

Due to the environmental constraints and climatic aberrations, the social vulnerability of the marginal and small farmers and the poor people in rural area is accentuated. Even large farmers in the Telangana region are fleeing to urban centres because of their perception of threat from the prevailing extremists movement in rural areas. This may be seen as a warning of the consequences of unsustainable agricultural in the drought-prone areas in the state.

One of the main reasons for the emergence of the naxalite problem is the relentless sub-division and fragmentation of land holdings in the region. Since the 1970s, the legislative measures have become relatively sharper, and political mobilisation was also much in evidence, which resulted in the naxalite movement gaining currency in the countryside. One of the impacts of this was the decline in the net sown area and an increase in the extent of fallow land.

The proportion of marginal and smallholdings, which was about 57% in 1956 when the state of Andhra Pradesh was reorganised, increased to 83% by 2000–2001. In contrast, the proportion of large holdings declined from about 9% in 1956 to less than 0.5% from 2000 to 2001. During the same period, 95% of land holdings accounting for 73% of the area were less than 4 ha. There has also been substantial landlessness in rural Andhra Pradesh, and according to the National Sample Survey (NSSO, 2006), Andhra Pradesh has the second highest number of landless people after Punjab.

The large presence of agricultural workers also seems to indicate a more concentrated land distribution, which in turn is associated with a larger share of chronically dis-empowered people who lack the resources and the incentive to make the necessary productivity-enhancing investments.

In spite of considerable investment in irrigation resources by the state until the 1980s, there are several factors that have made them unsustainable. Canal and tank irrigation is declining due to lack of proper operational and maintenance expenditure, resulting in silting and degradation of catchment areas.

Farmers in the region bear a substantial burden of irrigation provision from their own resources – often by heavy borrowing – and still end up with unstable and low levels of irrigation. The disparity in the distribution of irrigation waters is further enhanced by social ostracism in the form of caste in the region and political dominance in the villages. A few influential farmers in the village often corner the benefits of development programmes, as the provisions of the *Panchayati Raj* – governance by people themselves at the village level – are taken for personal benefits. As a result, the rich farmers within the watershed are converting the community resource into private property and are cornering the harvested water for themselves.

One of the important factors contributing to agricultural growth is that the diversification from low value crops to high value and high productivity crops, such as maize, pulses, oilseeds and cotton, showed an increase. Ground nut in moisture-deficient Rayalaseema and highly vulnerable hybrid cotton in Telangana have exposed farmers in these regions to high-risk and unstable incomes, not only in the local but also in the global market. The recent NSSO report (2006) shows that incidence of indebtedness among farmers of Andhra Pradesh is the highest at 82%.

In addition to the rising cost of debt and other agricultural inputs with volatile commodity prices, the farmers in the state, particularly those who have been opting for high value crops like cotton, face serious problems of spurious and unregulated seed supply, undue reliance on disproportionate use of pesticides and due to lack of proper information they face a high risk of losing crops.

Trade liberalisation is another blow to the small and marginal farmers. The heavy burden of debt is perhaps the most important cause of agricultural distress, forcing thousands of helpless farmers across the rainfed region to commit suicide (Patil, 2008). This has forced the ruling Congress-led government to announce a loan-waiver of USD 1.33 million.

Recent initiatives, like the setting up of a Special Economic Zone (SEZ) in the country, have threatened to divest farmers of their land holding. This has led to social unrest that has been politicised for short-term gains (Bandyopadhyay, 2008).

As the government is moving in the direction of corporatisation of agriculture, it is not paying due attention to the development of small-scale irrigation schemes which could have created enormous rural employment. According to a study by Reddy (2006), in Andhra Pradesh small-scale irrigation alone can create 90 million person-days of employment with food security.

To address these issues, the present government has developed some schemes like the National Rural Employment Guarantee Act (NREGA) in 2005 by the Ministry of Rural Development. This has been named 'the people's act', as it aims to provide 100 days of unskilled employment in a financial year with the prime objectives of generating productive assets, protecting the environment, empowering rural women, reducing rural–urban migration and fostering social equity in society. Moreover, the government has floated several other schemes to reduce the stress and vulnerability of the rural people by introducing schemes like *Swarnajayanti Gram Swarozgar Yojana* (SGSY – Golden Jubilee village self-employment scheme), mid-day meal scheme, *Sarva Shiksha Abhiyan* (SSA – Universal self-education programme), which is deemed to be the largest programme of its kind in the world, and the Assured Rural Water Supply Programme (ARWSP) which was renamed as the Rajiv Gandhi National Drinking Water Mission in 1991 (NSSO, 2006).

7 Conclusion

Seventy one million persons are involved in agriculture in the rainfed states of peninsular India alone, which contributes a large proportion of food grains, pulses and oilseeds to the national food stock. The Watershed-Based Development Programme has been adopted as a major development initiative for improving rural livelihoods in the region. However, there has been an exodus from rural areas to neighbouring urban centres. In Andhra Pradesh, there has been an increase in migration since 2001 due to deterioration in agriculture and loss of sources of livelihood in the rural areas. To address this problem, the government has initiated a number of rural employment schemes to counter the agricultural workers' flight to urban areas. Unfortunately, most of these schemes have had little impact, owing to improper implementation. This is leading to an increase of rural violence that has highlighted the lack of access of the poor to land, water and forest. Political complications of coalition governments also add to the prevailing condition, making it impossible to implement the programmes in the right spirit, so that the maximum benefit for which the programmes were intended accrue to the beneficiaries.

Acknowledgements The authors are grateful to the ICAR for funding the study and Director CRIDA for facilitating it. The authors are grateful to the Osmania University authorities for encouraging the study. They further acknowledge the assistance provided by the project staff associated with the authors.

References

Bandyopadhyay, D. (2008): Does land still matter? *Economic & Political Weekly*, Vol. XLIII No. 5, March 8–14. Available from http://www.epw.org.in

CRIDA. (2006): Annual Report (2005–06): Central Research Institute for Dryland Agriculture, Santoshnagar, Hyderabad – 500 059. pp. 8.

CRIDA. (2007): Annual Report (2006–07): Central Research Institute for Dryland Agriculture, Santoshnagar, Hyderabad – 500 059. pp. 8.

DCO. (2001): Census of India 2001 – Andhra Pradesh. D – Series migration tables. Data Dissemination Unit, Directorate of Census Operations, Hyderabad, Andhra Pradesh.

DES. (2006). Statistical Abstract of Andhra Pradesh – 2006. Directorate of Economics and Statistics, Government of Andhra Pradesh, Hyderabad.

DoAC. (2003): Farm produce prices, minimum support prices (as on 1-5-03) according to Crops Yearbook. Department of Agriculture & Cooperation; Government of AP, Hyderabad.

DoAC. (2004): Agricultural statistics at a glance. Area, production and yield of principal crops, Department of Agriculture & Cooperation; Government of AP, Hyderabad.

Hanumantha Rao, C.H. (2000): Watershed development in India – Recent experience and emerging issues. *Economic & Political Weekly,* November 4, 2000. pp. 3943–3947.

Joshi, P.K., A.K. Jha, S.P. Wani, L. Joshi, and R.L. Shiyani (2005): Meta-analysis to assess impact of watershed program and peoples's participation. *Comprehensive Assessment Research Report 8*. Comprehensive Assessment Secretariat, Colombo, Sri Lanka. 17 p.

Katyal, J.C., K. Ramachandran, B.V. Ramana Rao, and M. Narayan Reddy (1993): Drought in India – An historical perspective. In: *Rainfed Agriculture Research Newsletter* (January 1993–July 1993): Nos. 2&3 Central Research Institute for Dryland Agriculture, Santhoshnagar, Hyderabad 500 659, India. pp. 18–28.

Katyal, J.C., K. Ramachandran, M. Narayana Reddy, and C.A. Rama Rao. (1996): Indian agriculture – Profile of land resources, crop performances and prospects. In: *Regional Land Cover Changes, Sustainable Agriculture and their Interactions with Global Change* (with focus on South Asian countries) Proceedings of International Workshop of COSTED – ICSU – UNESCO – IBN, Chennai, India, 16–19 December. pp. 16–34

Kerr, J., G. Pangare, and V.L. Pangare (2002): Watershed development projects in India: An evaluation. *Research Report – 127*, IFPRI, Washington, DC.

MoF. (2008): Economic survey (2007–2008) Ministry of Finance, Government of India. Available from http://indiabudget.nic.in/es2007-08 /esmain.htm

NSSO. (2006): Some aspects of operational land holdings in India (2002–2003), NSS 59th Round (January–December 2003), National Sample Survey Organization, Ministry of Statistics & Programming Implementation, Government of India, August 2006.

Springate-Baginski, O., V. Ratna Reddy, M. Gopinath Reddy, and S. Galab (2004): Watershed development in AP – A policy review. CESS (Hyd) & DFID pub.

Patil, B.V. (2008): Agricultural indebtedness: Crisis and revival. *Economic & Political Weekly*, Vol. XLIII. No. 5, March 8–14, Available from http://www.epw.org.in

Planning Commission. (2007): Report the working group on natural resource management eleventh five-year plan (2007–2012). Planning Commission, Government of India, New Delhi.

Ramachandran, K. (2001): Inventorisation of contaminated sites in India using remote sensing, GIS & conventional methods (April 2000–June 2001): MOEF – NGRI – CRIDA Project, New Delhi.

Ramachandran, K., and S. Padmaja (2005): Evaluating ecological sustainability of cultural landscapes in Southern India. The Jahangirnagar review, Part II: Social science. Vol. XXVIII, 2004, Printed in June, ISSN 1682–7422. Jahangirnagar University, Savar, Dhaka – 1342, Bangladesh. pp. 1–14.

Ramachandran, K., M. Gayatri, V. Bhasker, G. Srinivas, K. Venkatravamma, T. Srinivas, and M. Sankar Rao. (2007): Evaluating livelihood security of farm household in treated watershed in semi-arid region. *Indian Journal of Dryland Agricultural Research & Development* 22(1), 95–108.

Ratna Reddy, V., M. Gopinath Reddy, Y.V. Malla Reddy, and J. Sousson (2004): Sustaining rural livelihood in fragile environments: Resource endowments or policy interventions? *Working Paper No. 58*, CESS, Begumpet, Hyderabad, India. 33p.

Reddy, D.N. (2006): Half a century of travails of agriculture in Andhra Pradesh. In: Rao, R.S., Hanumantha Rao, V., and Venugopal, N. (eds.), *Fifty Years of Andhra Pradesh 1956–2006*. Centre for Documentation, Research and Communication, Hyderabad. pp. 137–161.

Shah, A., R. Devlal, H. Joshi, J. Desai, and R. Shenoy (2004): Benchmark survey for impact assessment of participatory watershed development projects in India. Report submitted by GIDR, Ahmedabad to Planning Commission, Government of India, New Delhi. 117 p.

Singh, H.P., B. Venkateswarlu, K.P.R. Vittal, and K. Ramachandran. (2000): Management of rainfed agro-ecosystem. In: Yadav, J.S.P., and Singh, G.B. (eds.), *Natural Resource Management for Agricultural Production in India*. International Conference on managing Natural Resources for Sustainable Agricultural Production in the 21st Century February 14–18, New Delhi. 1022p.

Sreedevi, T.K., B. Shiferaw, and S.P. Wani (2004): Adarsha watershed in Kothapally: Understanding the drivers of higher impact. Global Theme on Agro-ecosystem Report No. 10, International Crops Research Institute for the Semi-arid Tropics, Patancheruvu – 502 324, Andhra Pradesh, India. 24p.

Velayutham, M., D.K. Mandal, C. Mandal, and J. Sehgal (1999): Agro-ecological Subregions of India for Planning and Development. NBSS Publ. 35, NBSS&LUP, Nagpur, India. 372p.

Vittalmurthy, K.P.R., M.V. Subrahmanyam, and T. Satyanarayana (2008): Tele-connections in Indian Monsoon. In: *Conference Proceedings of International Symposium on Agrometerology and Food Security*, 18–21 February, 2008, CRIDA, Hyderabad, pp. 88.

Chapter 10
Migration to Contaminated Sites: Migrants' Settlements in Central and Eastern Europe Built in Places with High Environmental and Social Vulnerability

Richard Filčák

1 Introduction

Porto Romano, on the outskirts of the Albanian port of Durres on the Adriatic Coast, is an old leather-tanning and pesticide-producing complex abandoned since 1990. Following the collapse of the totalitarian regime, waves of rural migrants from northeast Albania looking for economic prosperity settled in the area. In the late 1990s, refugees of the Kosovo conflict also made their home here. Rudňany is a village situated in the eastern part of the Slovak Republic. Migrant workers from the Roma ethnic minority attracted by the 1960s boom in the metal-processing industry settled down on the outskirts of the village.

What do these two distinctive places have in common? In both of them, migrants (voluntarily or involuntarily) found themselves living in places with highly contaminated land, surrounded by piles of industrial waste and endangered by security and health threats. There is a growing literature dealing with environmental causes of migration. However, there is less exploration of the environmental factors that determine the settlement decision of the migrants. People who left their home because of war, ethnic clashes, social pressures, or environmental pollution and disasters may find that their new settlements are safe from some of the stressors (e.g., ethnic violence) but may face environmental threats (e.g., soil or water contamination).[1]

People are not equal in the access to safe locations for settlement. Social, economic, and environmental factors play an important role in the selection process. Migrants are often vulnerable. They do not possess resources, social capital, or networks and have limited bargaining power over selection of a place

R. Filčák (✉)
Institute for Forecasting Studies, Slovak Academy of Sciences, Bratislava, Slovak Republic
e-mail: rfilcak@yahoo.co.uk

[1] I specifically focus in this chapter on migrants with weaker economic background, belonging to marginalized groups or ethnic minorities.

T. Afifi, J. Jäger (eds.), *Environment, Forced Migration and Social Vulnerability*,
DOI 10.1007/978-3-642-12416-7_10,

for settlement. Those possessing financial or social capital have choices. The others must take what is available or what they receive.

On the macro-level, their decisions often depend on borders (e.g., migration within the territories of the nation states or emigration to other countries with open or semi-open borders). On the micro-level, the concrete decision – where these people are directed to or allowed to settle down – is dependent on multilayered and interlinked social, economic, and environmental factors. Paradoxically, people migrating because of social reasons may find themselves endangered by environmental threats.

I argue in this chapter that the selection of places where migrants settle down is not accidental. It is rather the outcome of a complicated social process. Migrants – as a group with a weaker position in decision-making – are vulnerable and may face different ethnic or social prejudices or be exposed to open discrimination. Environment can play an important – if not crucial – role in the selection of settlements. I demonstrate these points for the cases such as Durres in Albania, Rudňany in Slovakia, and Mitrovica in Kosovo. When compared to the local people, the migrants or refugees are (or were) exposed unequally to the impact of the polluted environment, regardless of whether they have escaped to another state or migrated for work within nation–state territory.

Migrants and refugees are extremely vulnerable. They often lose everything they have possessed and look for temporary or permanent settlement in their new destinations. Environmental conditions are important factors in the decision-making. I argue that adverse environmental characteristics in some of the new migrants' settlements are the outcome of discrimination, marginalization, or economic interests and could be avoided. In other words, there is usually an alternative for the solution as defined by Pareto in his Pareto improvement (see, for example, Pareto, 1976; Powers, 1987): A change that can make at least one individual better off, without making any other individual worse off.

If houses for a migrants' or refugee camp are not built in an abandoned and contaminated industrial zone but in a safer area, nobody may be worse off, but lots of people may be better off. If refugees gain equal access to potable water, it is not decreasing anybody else's access to the water in a country that possesses enough of this natural resource. I argue that the migrants and refugees must be compensated for their lack of social capital and vulnerability in their new destinations by better involvement of the state, local authorities, and other stakeholders in enforcement of law and supporting equality in access to a safe environment.

2 Migration and the Environmental Vulnerability

We may define vulnerability as a function of exposure, sensitivity to impacts, and the ability or lack of ability to cope or adapt, while the exposure can be to hazards such as drought, conflict, or extreme price fluctuations, and also to underlying socio-economic, institutional, and environmental conditions (UNEP, 2007). Migrants are people at risk and their vulnerability depends on two factors: their sensitivity to the exposed conditions and the ability to cope or adapt.

It goes without saying that it is difficult to generalize a heterogeneous group of people such as migrants. In this chapter, I focus on migrants who are – because of ethnic affiliation, class, or cultural differences – discriminated against or marginalized in the places they find for settlement. There are many types of settlements where the poor migrants seek places to settle down. They are often characterized as an amalgam of overcrowding, poor, or informal housing, insecurity of tenure, unclear property rights, problematic residence permits, and inadequate access to safe water and sanitation.

The other common feature is the location on the outskirts of big cities or on the peripheries of villages. As Davis (2004) points out: "The poor [...] are everywhere forced to settle on hazardous and otherwise unbuildable terrains – over steep hill slopes, riverbanks and floodplains. Likewise, they squat in the deadly shadows of refineries, chemical factories, toxic dumps, or in the margins of railroads and highways." On the other hand, newcomers are often not only victims but also contributors to the adverse environmental change and increasing human vulnerability.

Large-scale re-settlement of people imposes additional environmental burdens on the areas to which they move. As Black and Sessay (1998) point out in their analytical framework: (1) refugees in a zone increase population/resource ratios, and resource accounting must consider the extent of renewable resources, the use of stocks of fixed capital, and the extent to which resource use generates technological or socio-economic changes that influence the ratios and geographic area; (2) refugees tend to be "exceptional resource degraders[2]"; and (3) refugees may ignore or be excluded from sustainable resource-use regulations. As I show for the Central and Eastern European cases, the last point of the analytical framework (i.e., exclusion from sustainable resource use) is especially valid in the selection of settlements for migrants.

3 Migrants and the Contaminated Sites

In this section, I demonstrate for concrete cases some of the theoretical arguments I outlined above. The cases then serve for analyzing the social processes leading to unequal exposure of migrants to environmental threats and for better analyzing the complex problem of migration vis-à-vis environment. The three examples selected in Central and Eastern Europe are Durres in Albania, Mitrovica in Kosovo, and Rudňany in the Slovak Republic.

[2] However, this is not necessarily a pattern followed after every forced migration. Black and Sessay (1997) examined the environmental consequences of forced migration in the Matam and Podor regions of the Middle Senegal River Valley. They compared wood fuel use by refugees and local populations. Drawing on a household survey and direct measurement of wood fuel use, little or no evidence was found to support the expectation that refugees use more wood for fuel than local people do, or that they are more destructive in their collection or use of wood.

3.1 Durres, Porto Romano in Albania

The abandoned industrial site on the outskirts of the Albanian port of Durres is an old leather-tanning and pesticide-producing complex closed since 1990. After the fall of the totalitarian regime that restricted in-country migration, rural migrants from northeast Albania looking for economic prosperity settled in the area. The late 1990s brought refugees of the Kosovo conflict, increasing the number of settled migrants to an estimated 6,000–10,000 (see, for instance, Brown, 2003; UNEP, 2000). Most recent results of the 2008 study conducted by Landell Mills for the United Nations Development Program (UNDP) and the Government of Albania (co-financed by Delegation of the European Commission to Albania) came to the conclusion that there are currently some 5,000–7,000 migrants settled on the flanks of Durres hills and on Durres plain. No recent and actual population data are available for the area of Porto Romano since a large portion of the settlements there is reportedly unregistered. There is also a significant number of people (primarily from the Roma minority) scavenging on the waste dump (Landell Mills, 2008). The abandoned chemical plants provided the squatters with building materials for new homes and access to water and electricity. The ground they built on, the bricks they used, and their water supply were all heavily contaminated with toxic chemicals, such as lindane, a banned nerve agent and carcinogen, and chromium, a chemical used in tanning known to cause kidney and liver damage and lung cancer. Additionally, according to tests made by a United Nations Environment Program (UNEP) team in 2000, a groundwater sample from a well in Porto Romano demonstrated 4.4 mg/l of chlorobenzene, over 4,000 times the acceptable level for drinking water in some EU nations (UNEP, 2000). High levels of chlorobenzene can damage the liver and kidneys and affect the brain.

The pollutants are so pervasive that livestock grazing upon the grass in the area has been contaminated. Soil samples revealed that there were high hexachlorocyclohexane (HCH) isomer concentrations, in the range of 1,290–3,140 mg/kg. In some parts of the EU, remediation is required when concentrations reach 2 mg/kg. The area is so toxic that in 2001, UNEP declared it as a "disaster area."

After initial hesitation, the Albanian government began remediation. It is unclear on what the extent of the cleanup will be. Previous efforts in 2002 led by the UNEP and the World Bank were hampered by squatters who refused to vacate the area, because they had no other place to go. An atmosphere of distrust between local communities and refugees has prevented effective mobilization to clean up the site (Schnabel, 2001). However, relocating squatters living directly within the old factory complex – the most toxic inhabited area in Porto Romano – and sealing them off is a high priority in the remediation effort.

3.2 *Rudňany, the Slovak Republic*

Rudňany is a village situated in the eastern part of the Slovak Republic. Migrant workers from the Roma ethnic minority attracted by the 1960s boom in the metal-processing industry originally settled down in the center of the village, but were later forced to move into two settlements (separated from the main village) at Zabíjanec and Pätoracké. The Roma ethnic minority faces prejudices and discrimination from the Slovak majority. Both settlements face serious environmental problems, which were present and documented well before the migrants settled down.

Mining activities from the past started to endanger Pätoracké due to subsidence. There are several kilometers of mine corridors right under the settlement. Sources of pollution in Zabíjanec are twofold. The first one is the surrounding area with toxic dumps of mining waste. Since these dumps are on the slope above the settlement, when it rains and snow melts, a high content of heavy metals flows down into the settlement and the heavy metals are deposited in the soil (Koči et al., 1994; Ochodnicky, 1996). Inhabitants then may bring this contaminated soil into their houses on their feet or inhale it in the dust during the summer. The second source of contamination is the place itself. Since it served as an industrial zone, there are residues of heavy metals, oils, and other industrial materials in the soil.

The metal-processing factory and mine gradually went bankrupt in the 1990s. More than 35 years after the people who migrated to the village settled down in these places, they remain endangered by the local environmental threats. The local municipality in collaboration with NGOs started a re-settlement program in 2002 for moving the people out of the two settlements, but the process is going extremely slowly. Moreover, the place for building new houses for the re-settled people is again in close proximity to the mine waste dumps and detached from the village.

3.3 *Internal Displacement Camps in Mitrovica, Kosovo*

When in 2005 a 2-year-old child died as a direct result of exposure to lead in the soil, the camp had already a history of 6 years of administration by the UN Interim Administration Mission in Kosovo (UNMIK) and the Provisional Institutions of Self-Government (PISG).[3] Before they were finally closed in September 2006, the camps known as Chesmin Lug, Zitkovac, and Kablare were situated in a highly toxic area due to tailings of the nearby Trepca Mine and Smelter Complex.

[3] The original plan was to keep the camps as a temporary arrangement to last only 45 days, but it took 6 years of concentrated advocacy and lobbying from residents, local, and international organizations to re-settle the migrants. All this happened despite growing evidence that living conditions in these camps were dangerous.

The area hosted around 500–600 internally displaced persons. The story started to unfold in 1999, after the Yugoslav army withdrew from the Kosovo territory and the NATO-led bombing ceased. In many places ethnic Albanians started their own revenge on Serbs and also included into the ethnic cleansing – to a prevailing extent – neutral Roma. The latter traditionally face ethnic discrimination from Serbs and Albanians in Kosovo, and Albanians used the opportunity to push them out from their territory in the turmoil after the conflict. The camps were closed in 2006 and migrants were moved to Osterode, as a result of the concentrated efforts of UNMIK and the UN High Commissioner for Refugees (UNHCR), the World Health Organization (WHO), the UN Children's Fund, and NGOs.

Steger et al. (2007) analyzed the case, and besides the role of the military logistics and environmental ignorance, they point to the role of the local Albanian administration in the place selection. As the UN envoy in Kosovo, Marek Nowicki, remarks, "These people were being treated this way for no other reason than that they were Roma."[4] The camps illustrate that environmental conditions may play an important role in generating human vulnerability, and ethnic discrimination may also involve discrimination in access to a safe environment.

4 Selection of Settlements as the Outcome of Social Processes

The cases above illustrate the vulnerability of migrants to the environmental conditions. However, why do people end up in such locations? What are the social processes leading to the migration of people and their settlement in places where local residents would never build a house? What role do the environmental conditions play in the selection process when the majority decides where these – often unwelcome – migrants can or must live? Is it an accident that migrants settle down next to a landfill on contaminated land or that they are regularly exposed to floods? Central and Eastern European experience shows that there was – and, unfortunately, still is – discrimination in equal access to environmental benefits (i.e., safe place for settlement) and people are not equal when it comes to exposure to adverse environmental impacts.

4.1 Exogenous and Endogenous Factors in Allocation of Space

Treatment of migrants and their settlement is an outcome of complicated social processes, where several factors enter into the picture. We may divide them into exogenous and endogenous factors. The former can be described as: (i) social; (ii)

[4] Hajdari, Ismet. Kosovo Roma refugees face unspeakable poverty, exposure to toxic metals. Agence France-Presse (AFP), April 10, 2005.

EXOGENOUS - SOCIAL - Class - Ethnic - Religion - Past tensions/disputes (e.g., over land or resources) - Cultural values (e.g. urban vs. rural)	EXOGENOUS - ENVIRONMENTAL - access to water, natural resources, - level and spatial distribution of contamination - „beyond the pale" as the socio-environmental syndrome
EXOGENOUS - ECONOMIC - price of the land (may reflect level of contamination or other environmental threats) - commercial potential of the land (e.g., tourism) - Competition on the job market	ENDOGENOUS - lack of social capital - lack of financial capital - cultural/language barriers - time stress: Maslow theory of needs[5] - lack of information: asymmetric information theory

Fig. 10.1 Exogenous and endogenous factors in the settlements location decisions (source: the author)

economic; and (iii) environmental, and they practically describe conditions under which newcomers enter. The latter are related to an internal capacity and capabilities of the migrants to cope with the exogenous conditions. See Fig. 10.1.

Social factors like class (e.g., poverty), different ethnicity, or religion, history of the past conflicts, or simply different cultural values may work against integration of the migrants and a fair selection of settlement areas. The outcome is an effort to segregate the newcomers from the community and create spatial distance from their places.

Gordon Allport (1954) identified a five-point scale for different intensities of prejudices and discrimination: (1) gossiping, stressing only negative aspects, or stereotyped estimations in which the members of a group or nation are represented as lazy, cunning, cowardly, dirty, evil, etc.; (2) avoiding any contact and creating a social distance; (3) discrimination in various spheres of life by which particular rights of the group in an inferior position are not respected; (4) physical attacks as a transition from the verbal to bodily aggression; and (5) extermination (pogroms, genocide, ethnocide). Prejudices and discrimination may play a role in allocation of space for migrants. Negative stereotypes of the newcomers are reflected in the tendency to allocate places for them outside of the settlements to avoid direct contact. Discrimination in different aspects of life (e.g., access to pastures, water, or school system) may lead to escalation of tensions.

Prejudices and discrimination influence economic decisions. A purely "homo economicus" perspective would be to those of the communities and/or individuals to sell land if the price is high enough and the place is of no specific importance to them. In reality, we see that the decision of groups and individuals in these communities is influenced by social and cultural factors; people would rather not sell or provide land for a refugee camp or allow newcomers considered to be "inferior" or "dangerous" to live in close proximity. In Rudňany, Slovakia people refused to sell land for new social houses for the Roma, even if they were in a difficult social situation themselves and needed the money. This was partly because they did not want the people in the proximity of their dwellings, partly because they were afraid of the reaction from their own community.[5]

An important aspect related to economic interests is to protect the local job market and wages, which may play against migrants. For instance, separating newcomers from the centers and/or transport infrastructure may disadvantage newcomers and protect the jobs of the old settlers. Economic interests may also force communities to hide camps or settlements of migrants, since they believe that these places may negatively affect the tourist industry or investments.

Authorities building new camps for IDPs[6] like those in Mitrovica or migrants seeking land to buy or rent from local people may face the situation which we can describe with the theory of asymmetric information.[7] It suggests that markets do not always operate efficiently, because buyers and sellers do not always have access to the information they need to make optimal choices. This means inefficiency in markets that people had not thought of before, namely, imperfect information. Sellers (or people identifying land for temporary or permanent settlements) have a comparative advantage over the people seeking places for migrants.

When a company wants to buy land or another company, it usually goes through the process of environmental audit of the potential acquisition. Its aim is to address the problem of the asymmetric information and reflect potential environmental burdens in the price. This is hardly the case when a state, international agencies, or NGOs need to build a camp for refugees or when migrants want to find a place for housing.

In the case of hostile attitudes to migrants, the environmentally problematic places are sometimes those which are allocated, sold, or rented to these migrants. In other cases it is just coincidence that the environmentally problematic places are cheap, useless, or distant enough from the settlements. In other words, we may see prejudices, discrimination, and social and economic interests of the side of the majority, and vulnerability, lack of financial and social capital, and lack of information on the side of the newcomers. The result then may turn out to be settlement in environmentally problematic spots in the form of abandoned industrial zones, regularly flooded areas, or places next to landfills.

4.2 Migrants and Their Territory as "Beyond the Pale" Space

People may be directed to settle in areas that are considered by the locals to be worthless or marginal. In other cases, the places may become so when they start to be inhabited by unwanted "intruders." Once you have space inhabited by

[5] Maslow established the theory of a hierarchy of needs, suggesting that human beings are motivated by unsatisfied needs and that certain lower needs need to be satisfied before higher needs can be satisfied. For instance, very basic needs such as air, water, food, or sleep have priority over safety needs (Maslow, 1970).

[6] Personal interviews done by the author, Summer 2005.

[7] Internally displaced persons.

people perceived by the majority as "inferior," it becomes – in the eyes of the majority, i.e., decision-makers – an unsuitable place for any development plan or project.[8] On the contrary, it becomes an area for the allocation of problematic projects and activities. Once the camps for migrants are set up, the area may become a place for environmentally problematic management practices (e.g., the location of a new landfill in close proximity). Parts of territory may start to be considered "lost" for any meaningful investment, and the local council is well aware of the fact that migrants – as the least powerful group on the local level – will not generate sufficient pressure to prevent this type of investment.

The area inhabited by migrants in the villages' or cities' territory may start to be constructed by the majority as a lost ground or territory "beyond the pale" (Filcak, 2007). The expression "beyond the pale" referred in the past to various defended enclosures of territory inside other countries. It meant "territory outside the bounds of acceptable behavior." The idea behind it was that civilization stopped at the boundary of "the pale" and beyond lay those who were not under civilized control and whose behavior therefore was not that of civilized people.

5 Conclusions

Environmental vulnerability may be a cause but – paradoxically – also an effect of migration. In other words, adverse environmental conditions in some cases may also be significant factors contributing to the vulnerability of migrants in places they find for new settlements. Competition and conflict between two groups of unequal social and political status over access to a safe living environment were then set forth as an explanatory framework for the migration analyses.

Migrants often lack financial and social capital and – as the "newcomers" to villages or cities – have limited possibilities to choose places for their settlements. In the cases where ethnic or cultural prejudices or discrimination occur, environmentally problematic areas are selected for settlements of migrants. The majority of the population may allow them to settle only in the places that are separated from their dwellings by natural or artificial barriers (e.g., river or railway) and/or which are of small commercial importance for them. As the cases described in this chapter demonstrate, these places may also be contaminated or unsuitable for life. The majority may also start with the conceptualization of the migrants' areas as "beyond the pale" spaces where environmentally controversial practices may be gradually concentrated.

---

[8] Theory of asymmetric information: George Akerlof, Michael Spence, and Joseph Stiglitz developed this theory challenging laissez-faire principles in setting prices in market economy (Nobel Prize in 2001).

Migrants or refugees should not become victims of the local environmental conditions and their exposure to the environmental threats should be equal and not discriminatory. It is important that environmental conditions are taken into account when refugee camps are planned or when social actors plan settlements for migrants. The state in collaboration with other regional and local authorities and in cooperation with international organizations and NGOs should protect law enforcement and prevent discriminatory approaches to migrants. A strong international and nation–state framework of human rights and equal opportunities must encompass also areas of the environmental vulnerability and provide guarantees that biases and prejudices against refugees and migrants do not put them into vulnerable positions, in order to enable them[9] to integrate into the society.

References

Allport, G.W. (1954): *The Nature of Prejudice*. Cambridge University Press, Cambridge, MA.

Black, R., and M. Sessay. (1997): Forced migration, environmental change and woodfuel issues in the Senegal River Valley. *Environmental Conservation* 24, 251–260.

Black, R., and M. Sessay. (1998): Forced migration, natural resource use and environmental change: The case of the Senegal River Valley. *International Journal of Population Geography* 4(1), 31–47.

Brown, V.J. (2003): Give me shelter: The global housing crisis. *Environmental Health Perspectives* 111(2), 92–98, February, 2003.

Davis, M. (2004): Planet of slums: Urban involution and the informal proletariat. *New Left Review* 26, 5–34, March/April.

Filcak, R. (2007): Environmental justice in the Slovak Republic: The case of Roma ethnic minority. Dissertation, Department of Environmental Sciences and Policy, Central European University, Budapest.

Koči, M. (1994): *Riešenie ekologickej situácie v oblasti Stredný Spiš za rok 1994: Záverečná správa* Addressing ecological situation in the Stredný Spiš region in 1994: Final Report. Košice: Slovak Academy of Sciences.

Mills, L. (2008): Rapid environmental assessment for an industrial park and an energetic park at Porto Romano, Durres. Report release on the 25 May, 2008.

Maslow, A. (1970): *Motivation and Personality*. Harper, New York, NY.

Ochodnicky, M. (1996): *Emisno-imisná situácia v zaťažených územiach oblastí Stredný Spiš a Jelšava-Lubeník – súčasný stav a predpokladaný vývoj do roku 1999*. [Emission-immission situation in the polluted areas of Stredny Spiš and Jelšava-Lubeník: The present situation and expected development until the 1999], Envitech, Trenčín.

Pareto, V. (1976): *Sociological Writings*. Rowman & Littlefield, Lanham, MD.

Powers, C. (1987): *Vilfredo Pareto (The Masters of Sociological Theory)*. SAGE Publications, London.

United Nations Environment Programme. (2000): UNEP Balkans post-conflict environmental assessment, Albania. Analytical Results of UNEP Field Samples from Industrial Hot Spots. UNEP. Geneva.

[9] This position can be described as ethnocentric, while the central group's perception is that their race, culture, and norms are superior to those of others. In this way, other groups and their individual members are judged as less valuable or important.

United Nations Environment Programme. (2007): Vulnerability of people and the environment – Challenges and opportunities. In *Global Environment Outlook (GEO-4): Chapter 7*. Environmental Information Coalition, National Council for Science and the Environment. Washington, DC.

Schnabel, A. (2001): *Southeast European Security – Threats, Responses, Challenges*. Nova Science, New York, NY.

Steger, T., A. Antypas, L. Atkins, F. Borthwick, C. Cahn, R. Filcak, K. Harper, I. Malbasic, and K. Medarova (2007): *Making the Case for Environmental Justice in Central & Eastern Europe*. CEU Center for Environmental Policy and Law (CEPL) and The Health and Environment Alliance (HEAL), Budapest.

Chapter 11
Migration and Natural Hazards: Is Relocation a Secondary Disaster or an Opportunity for Vulnerability Reduction?

Nishara Fernando, Koko Warner and Jörn Birkmann

1 Introduction

Natural hazards result in substantial human movement and displacement. As the world approaches the year 2050, scientists expect that tipping points will be exceeded and rapid-onset natural hazard events will increase (IPCC, 2007). These developments could move migration to a new magnitude (Christian Aid, 2007; IOM, 2008; Myers, 2002).

Recent cyclones in Burma and earthquakes in China have together displaced 17.4 million people temporarily. Within a period of 1 month in early 2008, millions of people were displaced by natural hazards – 15 million in Sichuan, China (IFRC, 2008) and an estimated 2.4 million in Myanmar (Lom, 2008). However, these snapshots taken during and directly after the disasters are not sufficient. For an improved understanding of the nexus between vulnerability and migration induced by environmental stressors, it is interesting to examine whether relocation is a secondary disaster or an opportunity for vulnerability reduction. Since the scale of such migration flows linked to environmental stressors is expected to rise with unprecedented impacts on lives and livelihoods, these questions need more attention.

Temporary displacement from natural catastrophes can further lead to permanent migration, as illustrated by the 2004 Indian Ocean Tsunami and the 2005 Hurricane Katrina. The Indian Ocean Tsunami in late 2004 displaced over 2 million people (AidWatch, 2006). The UN Office of the Special Envoy for Tsunami Recovery estimates that 1.5 million people lost their livelihoods in the aftermath of the tsunami, further complicating the resettlement of migrants (ibid). Another devastating natural catastrophe, Hurricane Katrina, resulted in the largest displacement of Americans in the country's history (see also, Chapter 3, by François Gemenne, this book). Hurricane Katrina ultimately caused about 1.5 million people to be displaced temporarily and an estimated 300,000 people permanently (Grier, 2005). Of the 1.5 million displaced people,

K. Warner (✉)
United Nations University Institute for Environment and Human Security, Section on Environmental Migration, Social Vulnerability, and Adaptation, Bonn, Germany
e-mail: warner@ehs.unu.edu

T. Afifi, J. Jäger (eds.), *Environment, Forced Migration and Social Vulnerability*,
DOI 10.1007/978-3-642-12416-7_11,

an estimated 107,000 illegal immigrants and temporary guest workers experienced secondary displacement due to Katrina – these affected people were already migrants when Katrina forced them to move again (Castillo, 2005). As the number of economic and other forms of migrants grows world-wide, the potential for displacing people who are already uprooted grows. This presents a new challenge for humanitarian assistance, emergency and evacuation planning, and post-disaster rehabilitation programmes.

1.1 How Many People Are Exposed to Migration and Displacement from Natural Hazards?

Numerous estimates of environmental migration fluxes have been published, and there is a growing consensus that migration will increase substantially in the future (e.g. Myers, 2005). Estimates of potential environmentally displaced people range widely, from 17 million (Leighton, 2006) to currently 24 million (UNHCR, 2006). The Office of the United Nations High Commissioner for Refugees (UNHCR, 2002), for example, estimates approximately 24 million people around the world who have fled because of floods, famine, and other environmental factors. The Almeria Statement (1994) mentioned that 135 million people could be at risk of being displaced as a consequence of severe desertification. In a 2002 paper by MEP Jean Lambert, a Green Party member of the European Parliament, the author estimated that the number of people displaced by climate change in China alone was 30 million (Lambert, 2002). Myers (2002, 2005) estimated that 25 million people in 1995 had migrated with a possible doubling of that number by 2010 to 50 million. A widely cited projection for the period leading up to the year 2050 is 200 million environmentally induced migrants (Brown, 2008). By 2050, one estimate reaches almost 700 million people on the move because of environmental factors or almost 1 in every 11 people living on the Earth at that time (Christian Aid, 2007).

All these estimates, including their underlying methods and assumptions, are subject to debate. Patterns of migration and displacement will evolve and intermingle differently with social, economic, and political factors, depending on whether the environmental hazard is a slow or fast-onset event (Renaud et al. 2007). The complexity of interactions makes reliable estimations of environmentally induced migrants challenging (Döös, 1997; Warner et al., 2008). Quantification is further complicated by the fact that environmentally induced migration is mostly internal (at least in the initial phase).

Few authors contest that environmentally induced migration falls much below 20 million environmentally induced migrants today. Even the most widely cited estimate of 200 million migrants by 2050 (Brown, 2008) suggests that environmentally induced migration could soon involve up to 3% of the current world population (CIA, 2008) in just four decades from now. The social and economic costs of this uprooting, accounting for both losses and responses, have not been estimated yet.

2 Global Assessment: Migration Induced by Environmental Stressors

The recent report of the German Advisory Council on Global Change (WBGU) underlines that drought, desertification, and other forms of water scarcity are estimated to affect as much as one-third of the world's human population and could contribute to people leaving the affected areas to secure their livelihoods. Figure 11.1 identifies hotspots where environmental problems, conflict, and migration pressures overlap. These are areas that may face serious environmentally induced migration in the future. The map indicates that arid and semi-arid areas manifest particular tendencies towards migration, conflict, and water and food shortages.

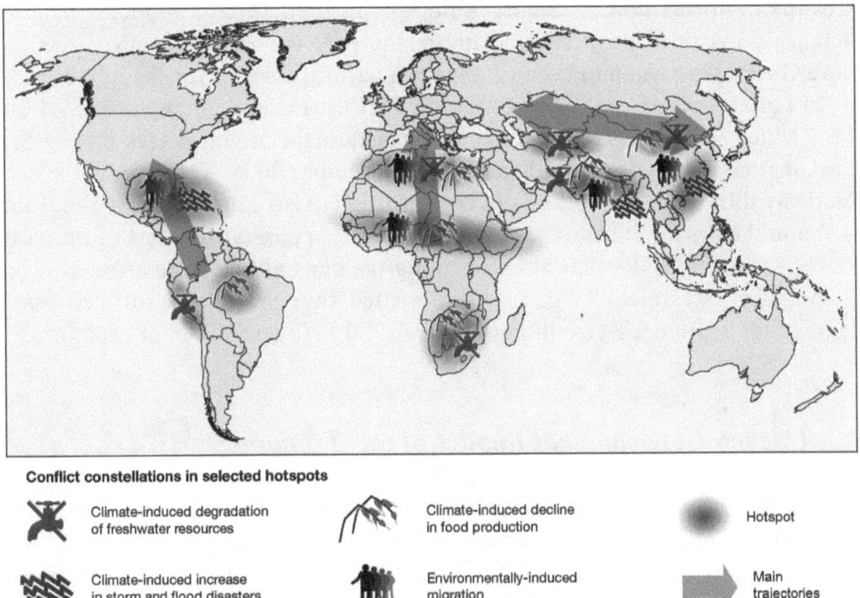

Fig. 11.1 A map of conflict and migration induced by environmental stressors (source: German Advisory Council on Global Change WBGU (2007): Climate Change as a Security Risk arrows added by UNU-EHS)

Research suggests that certain regions of the world will experience migration-related impacts of climate change more dramatically than others. The devastating impact of climate change will be most acute in regions and among groups that face multiple stressors at the same time – pre-existing conflict, poverty, and unequal access to resources, weak institutions, food insecurity, and incidence of disease. Research suggests that environmentally induced migration could be felt most dramatically in the already arid and semi-arid areas in developing countries around the globe – Africa, the Middle East, the mountainous areas of the Far East including India, Pakistan, and Afghanistan. Many of these areas already host large populations and struggle with conflict – often in hazard-prone areas.

The global projections also underline that environmentally induced migration could be a result of major stressors like floods and droughts.

3 Local Assessment: Migration and Relocation After a Disaster

3.1 Case Study: Sri Lanka – Relocation After the Tsunami

Interestingly, Sri Lanka too is prone to floods and droughts. However, also cyclones, landslides, and coastal erosion are also important environmental stressors. Various hazard events – floods, drought, cyclones, etc. – caused increasing losses of property and life in the past few decades (DMC, 2005). Nevertheless, the tsunami became a known natural disaster for the inhabitants of Sri Lanka only after the Indian Ocean Earthquake with a magnitude 9.1 on the Richter scale triggered a series of devastating tsunami waves hitting Sri Lankan coastal towns and villages on 26 December 2004. The tsunami struck relatively thin but long stretches over two-thirds of Sri Lanka's coast line from Jaffna in the north, the entire eastern and southern coasts, and part of the west coast; 13 out of 14 districts situated along the coastal belt were affected. The preliminary assessment of damages estimated that Sri Lanka suffered asset damages of around US$1 billion (4.5% of GDP) (Jayasuriya et al., 2005).

3.2 Uneven Geographical Impact of the Tsunami

The Eastern Province which was one of the provinces directly hit by the tsunami reported the highest number of deaths (14,354) and displaced persons (218,727), while the Southern (10,056 deaths and 159,105 displaced persons) and Northern Provinces (6,230 deaths and 64,067 displaced persons) reported the second and third highest numbers, respectively. However, Galle, as a single district of the Southern Province, reported the highest number of displaced persons (128,077) (Department of Census and Statistics, 2004). An estimated 150,000 people lost their main source of income. At least one-third of them were employed in the fisheries sector, 45% in tourism and related services, and 4–5% in agriculture and livestock (GOSL, 2005; IPS, 2005).

About 13% of the housing stocks in the administrative divisions along the coast were either completely or partially damaged. The housing damage pattern classified by provincial level shows a significant proportion of completely damaged or partially damaged (unusable) housing units reported from the Eastern Province (27%), followed by another 20% from the Southern, and 18% from the Northern Provinces. Moreover, most of the completely or partially damaged unusable houses were situated within 500 m of the coast (Department of Census and Statistics, 2004).

The Sri Lankan government was not prepared to confront such a high impact natural disaster with such extensive damage on its own, and therefore relied on both the domestic and international communities for various types of aid for relief, rehabilitation, and reconstruction work.

3.3 Forced Migration

After the mega-disaster had occurred, the Sri Lankan Government established new policies and projects. The best known among them are the development of a tsunami early warning system and the enforcement of a 'buffer zone' along the coast. A buffer zone (no-construction zone) had already existed before the tsunami but had not been fully enforced. As a result of the massive destruction caused by the tsunami, the government took a decision to reintroduce and fully enforce a buffer zone of 100 m in the south and south west, and a 200 m buffer zone in the north and the east, as the human and material losses were higher there than in the south (GOSL, 2005). The government argued that this buffer zone was needed in order to minimise human and material losses in the case of future tsunamis. However, the implementation of the buffer zone regulation caused various protests and controversies among affected and non-affected households living in this zone. Furthermore, the different treatment of the Southern and Northern and Eastern Provinces contributed to a high level of mistrust. As a consequence, tsunami-affected people who had been living in the buffer zone before the tsunami had no other choice than to relocate and settle down in dwellings allocated to them. As a result of this buffer zone restriction, people who lived in about 70,000 damaged housing units in the coastal belt were forced or had the chance to settle down in relocation settlements situated mostly far away from their previous dwellings, particularly in the densely populated south coastal region, due to the fact that their previous settlements lacked suitable adjoining land to build new housing. Most of those who had been living on fishing resisted being relocated to settlements far from the sea, due to the disruption that such a distance might cause to their income-earning activities, although most of them did not ultimately have any other place to live other than the new settlements. On the other hand, illegal settlers, irrespective of their income-earning activities, showed their willingness to settle down in new settlements, driven by the aspiration of escaping poverty by obtaining a legally accepted permanent place to live (Birkmann et al., 2007). However, it is important to examine how these groups adapted to the new social, economic, and environmental conditions in the new settlements.

It is against this background that our research makes an attempt to explore how tsunami-affected forced migrants, who had lived in the Galle city area before the tsunami and are now living in settlements relocated in mainly rural areas far away from the city (8–12 km), manage to adapt to the new social and economic environment.

3.4 The City of Galle

The city of Galle is the capital of the Southern Province and is situated 119 km south of the commercial city of Colombo with 90,934 inhabitants (Department of Census and Statistics, 2002). Galle is among the cities most affected by the tsunami. It consists of a harbour, a cement factory, hospitals, public schools, and commercial buildings, as well as a fort built by the Portuguese in the sixteenth century which has now been declared a World Heritage Site by UNESCO. The Galle Municipal Council governs the city which also belongs to the Galle Four Gravates Divisional Secretary Division.

Due to the difficulty in finding spare land in the congested Galle city area, the majority of the people who had lived in the buffer zone of Galle before the tsunami had to relocate to new housing settlements built in the Akmeemana Pradeshiya Sabha area which falls under the Akmeemana Divisional Secretary Division and is situated 8–12 km away from their previous dwellings.

3.5 Relocated Lives – Findings from Interviews with Displaced People

Empirical data were collected from displaced people of three relocated settlements situated far away from the Galle city area. Most of the interview respondents stated that the settlement as a whole is socially and economically marginalised from the outside world, not only because of the bad behaviour of some settlers (alcoholics, illicit liquor and cannabis trafficking, crime and violence, etc.) but also due to the nature of their occupations as fishermen and day labourers. This situation was elaborated further by one key informant in an interview with a government officer as follows: 'Those notorious thugs and underworld gang members used to live in the Galle city area before the tsunami struck and are now living in this settlement. They are the ones who create most of the problems here under the influence of alcohol. There are also a few heroin addicts among them. People living in the surrounding villages are scared of these people as some villagers feel uncomfortable passing by this settlement alone during the dark after a few theft incidents near to this settlement have been reported'. An old settler living in a nearby village said, 'Most of our villagers engage in rice and tea cultivation. The majority of people living in the tsunami village are fishermen. Now this neighbourhood is dominated by them. They should live somewhere closer to the sea which is the ideal place for them. This is not the right place for these people to stay. Even the value of our land has dropped after the construction of these settlements'.

Conflicts were reported between old and new settlers over access to free land in the area as well as community centres in the new settlements. As a result, a few newly resettled families who were involved in physical violence with old settlers have already left their new housing.

As to the question of the main reason for relocation, most of the displaced people reveal that fear of another tsunami and their decision to resettle was right as they do not have any other place to stay in the Galle Municipal Council area. Push and pull factors that reduce or increase vulnerability after the tsunami due to specific interventions undertaken such as the establishment of the buffer zone and the relocation policy are outlined in Fig. 11.2 (Birkmann, 2007:67). Interestingly, some of the squatters who did not choose to live closer to the sea might now be able to benefit from the relocation process by being able to acquire a land title and get out of chronic poverty (Birkmann and Fernando, 2008). However, recent field work indicates that particularly relocation sites far away from the city centre have disrupted social networks. In addition, they lack the physical and social infrastructure needed in a functioning settlement. As a point of interest, when evacuees had been first relocated into the new settlements, government officials and various politicians had pledged their full support to develop important infrastructure facilities such as access roads, street lights, and transport within the settlements as well as income-generating activities as a strategy to minimise their uncertainty. These pledges have,

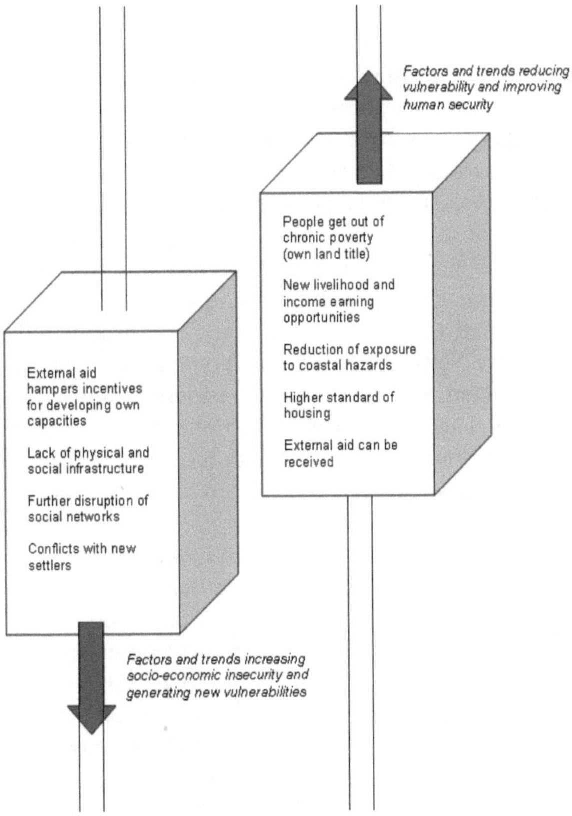

Fig. 11.2 Push and pull factors that reduce or increase vulnerability

however, only contrived to foster the hope in the minds of most of the displaced people that 'something good will happen to them in the near future'. When interviewed again 6 months after the first interviews had been conducted, the same people said that nothing had happened in the meantime in terms of developing basic infrastructure or creating income-earning activities other than repeating the same promises that the local politicians had made before.

The interviews also show that displaced people are not happy with the structure and the construction materials used to build their houses, as one can generally observe fractured walls, leaking roofs, and over-flowing drainages, although all the houses had been furnished with individual connections to water and electricity. Government officials conducted a survey covering most of these construction-related problems of the new settlements with the financial assistance from an International Non-Governmental Organisation. However, according to the local government officials, nothing has been done so far other than the above survey. At the time the interviews were held, none of the household heads had been given a legal document granting them the ownership of the land and the house they occupy. Most of the relocated dwellers do not even know who is responsible for granting them a legal title of land ownership.

Opportunities for income-earning activities in the new resettled areas are very limited, as most of the old dwellers are either engaged in rice or tea cultivation, possessing skills and land which new settlers do not have. Therefore, some new settlers have opened small-scale grocery shops in their houses as a strategy to improve their household income, as most of them are confronted with new expenditures, for example, for transport which they did not have in their previous dwellings. A few months after their opening, most of these shops had to close down, as they did not get enough business. Irrespective of these discouraging cases, most of the household members interviewed still aspire to start a small business in their houses, which would match with their skills, but they noted the difficulty in doing so without having the necessary financial capacity. Neither the donors who constructed the settlements nor the relevant government officials have considered improving the income-earning and income-generating activities of the new settlers after relocating them. The new settlers were only provided with a house. This pattern is confirmed further by other studies conducted in various relocation projects in Sri Lanka (Fernando, 2004; Hettige, et al., 2004).

It emerges clearly from the above evidence that most of the relocated people have difficulties in adapting to settlements situated far from the city, especially due to the lack of infrastructure, quality housing, employment opportunities, and the conflicts arising between newcomers and old settlers. This situation, in turn, forces some of the new settlers to leave their new settlements by renting, closing, or selling their houses, in order to either migrate to their previous places or other places closer to the city. Thus, for most of the tsunami-affected households which had been moved to new settlements, relocation did not trigger vulnerability reduction. This finding is particularly true for those who had worked in the fishing or the tourism sectors before the tsunami in 2004 occurred. For this group, the need for easier access to the work place soon became more important than the more remote threat of another tsunami which is a low-frequency hazard in Sri Lanka.

4 Conclusions

Livelihoods were destroyed and disrupted by the Indian Ocean Tsunami. Even though the majority of the affected households used to consider the buffer zone as an important strategy when established, the current situation of living far away from their place of work and the city adds further daily stress on households earning their living from the sea or tourism. Therefore, back migration into the buffer zone is currently a frequently observed phenomenon in Sri Lanka.

Furthermore, the two examples cited before, i.e. the global analysis of hotspots and the local assessment of forced migration after a disaster, show that natural hazards might trigger migration and relocation directly or indirectly.

The global assessment of the WGBU clearly gives emphasis to the influence of frequent hazards, such as floods and droughts as environmental stressors that contribute to migration. Interestingly, the global analysis views the hazard event itself as a major trigger and stressor. In contrast, the migration process set in motion after the low frequency but very devastating Indian Ocean Tsunami in Sri Lanka was largely the result of post-tsunami policy, i.e. the decision to relocate people from the most exposed coastal zones. However, the way relocation and resettlement programmes were conducted, the lack of integrative strategies, and the total underestimation of the dependency of households on coastal resources indicate the difficulties, problems, and shortcomings of the present relocation and planned migration strategies. On the other hand, it is worth noting that some of the households interviewed in the Southern Province of Sri Lanka have clearly stated that the buffer zone and the concept of relocation is a useful strategy. This is particularly true for squatters who hope to get a permanent place to stay through the relocation process.

5 Recommendations

Relocation needs careful planning as it intends to build up new communities by moving part of the population from one place to another. This implies changing the places where these people live, work, access job opportunities and education, build social networks, etc. Interestingly, most of the relocation settlements for the tsunami-affected forced migrants were built by various donors and

NGOs, according to the guidelines given by the Urban Development Authority and site plans developed by the National Housing Development Authority (NHDA). According to the Agreement between donors and the government of Sri Lanka, the donors pledged not only to build the houses in the settlements but also to bear the cost for establishing basic amenities (water, electricity, and sewage) in the settlement's housing and infrastructure (access roads, street lamps, community centres, etc.), while the government assumed the responsibility to develop the corresponding public services (Ministry of Finance, 2005). However, Hettige's (2007) study has revealed that tsunami-affected relocated households are now grappling with a range of more critical new issues such as the disruption of their traditional income-earning or income-generating activities, as the new settlements are situated far away from the coast, precarious access to services even below those they had in their previous dwellings, and the deterioration of long-established social networks. Therefore, while a majority of the relocated households might not be exposed to tsunamis again, other threats and risks to their daily life and livelihoods have become apparent. Further interviews are needed to validate the first results and to explore in more depth which households have really benefited from the relocation process in the longer run and for whom it implied a further disruption and destruction of their livelihoods.

Furthermore, it is important to note that global and local studies underline that many risk factors which may be individual or family-specific, such as illness, debt, or lack of livelihood opportunities – are directly linked to natural hazards and environmental conditions. The severity and extent of natural hazards affect migration. A gravity modelling exercise (Afifi and Warner, 2008), survey results, and expert interviews appear to indicate that the current environmental signals in migration are detectable. Events like tsunamis and floods or droughts can unleash processes that perpetuate existing migration. Migrant network connections link non-migrants, early migrants, and recent migrants through social ties. If environmental conditions worsen and people choose or are induced to migrate, it is possible that those remaining in areas facing environmental degradation will use the existing conduits to leave these areas. The theory and concepts of social-ecology, feedback-loop system, and coupled social–ecological system might provide insights into these dynamics. This means, they underline the necessity to capture feedback-loop processes and thus to apply a more system-oriented and process-oriented perspective. Additionally, the entitlement theory furnishes an interesting theoretical background with regard to the case study of Sri Lanka and the difficulties of relocated households to re-establish their livelihoods. Overall, we conclude that the influence of both frequent hazards such as floods and droughts and the influences and actual consequences of post-disaster policy – as illustrated in the case of Sri Lanka – need to be taken into account when examining the question of whether the phenomenon of relocation and migration is a secondary disaster or an opportunity for vulnerability reduction. Most studies still pay too much attention to the potential direct consequences of natural hazards, while we observe in many cases that actual migration and relocation is also heavily influenced by political decisions and strategies.

References

Afifi, T., and K. Warner (2008): The impact of environmental degradation on migration flows across countries. *Working Paper No. 5/2008; UNU-EHS Working Paper Series*. United Nations University Institute for Environment and Human Security (UNU-EHS), Bonn.

AidWatch. (2006): U.N. office of the special envoy for tsunami recovery. Reuters AlertNet. Accessed 24 December, 2006. Available from http://www.alertnet.org/printable.htm?URL=/db/crisisprofiles/SA_TID.htm

Almeria Statement. (1994): *The Almeria Statement on Desertification and Migration*, Statement following the International Symposium on Desertification and Migrations, Almeria, 8–11 February.

Brown, O. (2008): Migration and climate change. In: International Organization for Migration (IOM): *Research Series No. 31*, IOM, Geneva.

Birkmann, J., and N. Fernando. (2008): Measuring revealed and emergent vulnerabilities of coastal communities to tsunami in Sri Lanka: A prerequisite for sustainable reconstruction. *Disasters* 32, 82–105.

Birkmann, J., N. Fernando, S. Hettige, S. Amarasinghe, T. Jayasingam, D. Paranagama, A. Nandana, M. Naßl, S. Voigt, U. Grote, S. Engel, B. Schraven, B. Schraven, and J. Wolfertz (2007): Rapid vulnerability assessment in Sri Lanka, United Nations University Institute of Environment and Human Security (UNU-EHS), Bonn.

Birkmann, J. (2007): Tsunami: Socio-economic insecurities, direct and indirect impacts and the post-tsunami process: Special focus on Indonesia and Sri Lanka. Background Paper for World Economic and Social Survey 2008.

Castillo, E.E. (2005): Illegal immigrants afraid to get storm aid, Associated Press, 9 September, 2005.

Central Intelligence Agency (CIA). (2008): *The World Fact Book: World*. CIA, Washington, DC. Available from https://www.cia.gov/library/publications/the-world-factbook/geos/xx.html, 6 March 2008.

Christian Aid. (2007): *Human Tide: The Real Migration Crisis. Christian Aid Report May 2007*. London.

Department of Census and Statistics Sri Lanka. (2004): *Census of Buildings and Persons Affected by the Tsunami – 2004*. Department of Census and Statistics, Colombo.

Department of Census and Statistics. (2002–2003): *Area, Population, Registered Voters and Employees of Municipalities*. Department of Census and Statistics, Colombo.

Disaster Management Centre (DMC). (2005): *Towards a Safer Sri Lanka: Road Map for Disaster Risk Management*. Ministry of Disaster Management, Colombo.

Döös, B.R. (1997): Can large-scale environmental migrations be predicted?. *Global Environmental Change* 7, 41–61.

Fernando, N. (2004): A measure of urban vulnerable livelihoods – A case of Colombo. *Institute of Workers Education Journal* 4, 64–89.

German Advisory Council on Global Change (WBGU). (2007): *Climate Change as a Security Risk*. Earthscan, London. p. 4.

Government of Sri Lanka (GOSL). (2005): *Post Tsunami Recovery and Reconstruction: Progress Challenges and Way Forward*. Ministry of Finance and Planning, Colombo.

Grier, P. (2005): The great Katrina migration. In: *The Christian Science Monitor*. Available from http://www.csmonitor.com/2005/0912/p01s01-ussc.html, 12 September, 2005.

Hettige, S.T. (2007): Introduction. In: S.T. Hettige (ed.), *Tsunami Recovery in Sri Lanka: Retrospect and Prospect*. Social Policy and Research Centre (SPARC), University of Colombo, Colombo.

Hettige, S.T., N. Fernando, M. Mayer, C. Noe, and G. Nanayakkara. (2004): *Improving Livelihoods of the Urban Poor – A Study of Resettlements Schemes in Colombo*. German Technical Cooperation (GTZ), Colombo.

Intergovernmental Panel on Climate Change (IPCC). (2007): *Impacts, Adaptation and Vulnerability – Contribution of Working Group II to the Fourth Assessment Report of the Intergovernmental Report on Climate Change*. Cambridge University Press, New York, NY.

International Federation of Red Cross (IFRC). (2008): China: Sichuan Earthquake. Emergency appeal number MDRCN003. GLIDE number EQ-2008-000062-CHN. Operations update number 10, 6 June, 2008.

IOM. (2008): *Migration and Climate Change*, Report No. 31. IOM, Geneva.

Institute of Policy Studies (IPS). (2005): *Livelihood Post Tsunami: Build Back Better?* Colombo, Sri Lanka, 1–2 December, 2005, IPS, Colombo.

Jayasuriya, S., D. Weerakoon, N. Arunatilaka, and P. Steele. (2005): *Economic Challenges of Post Tsunami Reconstruction: Sri Lanka Two Years on*. IPS, Colombo.

Lambert, J. (2002): *Refugees and the Environment: The Forgotten Element of Sustainability*. The Greens/European Free Alliance in the European Parliament, Brussels, Belgium.

Leighton, M. (2006): Desertification and migration. In: Johnson, P.M., Mayrand, K., and Paquin, M. (eds.), *Governing Global Desertification*. Ashgate, London. pp. 43–58.

Lom, C. (2008): IOM Rushes Medical, Shelter Aid to Cyclone Nargis Survivors. IOM Feature Stories. Asia and Oceania. Wednesday, 02-07-2008. Available from http://www.iom.int/jahia/Jahia/media/feature-stories/featureArticleAS/cache/offonce?entryId=17705.

Myers, N. (2002): Environmental refugees: A growing phenomenon of the 21st century. *Philosophical Transactions of The Royal Society B* 357, 609–613.

Myers, N. (2005): *Environmental Refugees: An Emergent Security Issue*. 13th Economic Forum, Prague, 23–27 May.

Ministry of Finance. (2005): *Tsunami Housing Construction Programme: Homeowner Driven Housing Operational Instructions for Beneficiary Selection and Project Implementation*. Ministry of Finance, Colombo.

Renaud, F.G., J.J. Bogardi, O. Dun, and K. Warner. (2007): *Control, Adapt or Flee: How to Face Environmental Migration?* InterSecTions no. 5/2007, United Nations University Institute for Environment and Human Security, Bonn.

United Nations High Commissioner for Refugees (UNHCR). (2002): A critical time for the environment. *Refugees, No. 127*.

United Nations High Commissioner for Refugees (UNHCR). (2006): *Refugees by Numbers: 2006 Edition*. UNHCR/MRPI/B.1/ENG1, September 2006. UNHCR, Geneva. Available from http://www.unhcr.org/basics/BASICS/4523b0bb2.pdf, 12 March 2007.

Warner, K., T. Afifi, O. Dun, M. Stal, S. Schmidl, and J. Bogardi. (2008): Human security, climate change, and environmentally induced migration. In: Dokos, T. et al., (eds.), Climate Change: Addressing the Impact on Human Security. Hellenic Foundation for European and Foreign Policy (ELIAMEP) and Hellenic Ministry of Foreign Affairs, Athens, May 2008.

Chapter 12
Guatemala: A Review of Historic and Recent Relocation Processes Provoked by Disasters of Natural Origin

Yojana Miner Fuentes and Juan Carlos Villagrán de León

1 Guatemala: A Land Exposed to Many Natural Hazards

Due to its topography, geographic location at the intersection between three active tectonic plates and in the path of tropical hurricanes in the Caribbean region of the American hemisphere, Guatemala has been experiencing disasters throughout the centuries. As a product of the interaction between the Cocos, Caribbean, and North American plates, major earthquakes have destroyed communities and livelihoods; damaged or destroyed essential infrastructure for transportation, health, and education; and provoked economic losses, exceeding hundreds of millions of dollars. The earthquake on 4 February 1976 caused over 23,000 fatalities and affected over 5 million inhabitants throughout the country (OFDA-CRED, 2007). In a similar fashion, hurricanes have also caused disasters in floodplains and halted processes associated with development in urban and rural towns, leading to heavy impacts in the agricultural sector and on road infrastructure. The latest episode is hurricane Stan which occurred in September 2005 and led to major losses, only 7 years after hurricane Mitch impacted the rest of Central America in October and November 1998. Landslides, triggered either by intense rainfall or earthquakes, are feared due to their capacity to bury entire communities. Less frequent but equally destructive are eruptions of volcanoes such as Fuego and Santiaguito. Along the same lines, drought has also manifested itself, impacting rural communities, such as in the case of the drought of the year 2001 which forced the national government to declare a state of national calamity due to the geographical extent of the event (Government of Guatemala, 2001; Prensa Libre, 2001).

Y.M. Fuentes (✉)
Coordinadora Nacional para la Reducción de Desastres, CONRED, Guatemala, Guatemala
e-mail: yojaminer@yahoo.com

T. Afifi, J. Jäger (eds.), *Environment, Forced Migration and Social Vulnerability*,
DOI 10.1007/978-3-642-12416-7_12, © Springer-Verlag Berlin Heidelberg 2010

1.1 Historical Forced Displacements Provoked by Disasters in Guatemala

The first written accounts of forced displacements due to natural phenomena within Guatemala can be traced to the year 1541 during the period of the invasion by Spanish conquistadores. On 11 September 1541, torrential rainfall which could be associated with a tropical hurricane triggered a massive landslide that buried the town of Santiago de los Caballeros. At that time the latter was the location of the Spanish government in Central America (Juarros, 1936). The town was located on the foothills of the Agua volcano, and as a consequence of the disaster, the government decided to relocate the city and its survivors to the central part of the valley and further away from the slopes of the volcano.

Two centuries later, in July 1773, the new city of Santiago de los Caballeros was destroyed by the 'Santa Marta' earthquakes. As in the previous case, the Government relocated the city, forcing citizens to move to the new settlement, nearly 50 km away in the Valley of the Ermita, where the capital city of Guatemala stands today. The decision to relocate the city to this new valley took into consideration the view of placing a greater distance between such a new city and the hazard associated with volcanic eruptions related to Fuego Volcano which at that time was perceived as the root cause of such earthquakes and with respect to the Agua volcano in the case of landslides (Juarros, 1936).

In October 1762, the Village of Petapa was destroyed by a combination of a flood and a mudflow. Located on the floodplain of the Villalobos River, settlers never foresaw the risk of landslides triggered by rainfall occurring in the mountainous area of this basin. As a consequence of this disaster, the government established a new town called Villa Nueva (New Village), and relocated survivors to this new site, on a plateau far away from the floodplains of the Villa Lobos river (Juarros, 1936).

Examples such as these serve to illustrate the fact that new settlers often have no perception regarding natural hazards and only realise their exposure to such hazards through disasters. A similar statement can be made with respect to building materials and techniques. Again, new settlers may not be aware of the vulnerability of such materials and techniques with respect to particular hazards but only discover them through the destruction of infrastructure provoked by natural disasters, such as earthquakes.

The relocation processes in 1543 and 1762 arose as a result of the destruction of the villages or communities due to their exposure to hazards associated with landslides. In contrast, the relocation of Santiago de los Caballeros to its new site in the Valley of La Ermita in 1776 may have been ill-conceived by today's standards regarding seismic hazard. The exposure to earthquakes is nearly similar at both sites. However, the government at that time may have seen the relocation process as the only way to cope with such events.

A critical issue which is mentioned later in this chapter is the resettlement of citizens in destroyed towns. These resettlements took place decades or centuries later in three towns. The first town on Santiago de los Caballeros is currently a consolidated town and is called Ciudad Vieja (Old City). The second town of Santiago de los Caballeros is today called Antigua and is one of Guatemala's main tourist attractions. It is thus a source of income both to its citizens and to the government, in addition to the fact that it is one of UNESCO's cultural heritage centres. The third town is the old village of Petapa that for a while had adopted the name 'old town', but its name was changed nearly a century ago to Villa Canales. It is now a formal city that hosts the municipal government of the district of Villa Canales. The city is growing at a tremendous pace due to its location within the Guatemalan metropolitan area.

1.2 Recent Examples of Forced Displacements Provoked by Disasters in Guatemala

In 1983, the town of El Palmar, located in the foothills of the Santiaguito volcano, began to suffer from the impacts of 'lahars' propagating along the rivers Nima 1 and Nima 2. A torrent of sand, mud, and boulders, triggered by intense rainfall on the active cone of Santiaguito volcano, was channelled by these rivers and led to the destruction of one segment of the town. The sedimentation process associated with these lahars continued in subsequent years, until this town was literarily buried, forcing the government to declare a State of Calamity on 5 May 1987 (Government of Guatemala, 1987). Nearly a decade later, the National Coordinating Agency for Disaster Reduction (CONRED) declared El Palmar as a 'High-Risk Zone' on 5 June 1997, in order to prevent its former citizens from settling there again (CONRED, 1997). The new El Palmar village was inaugurated on 12 July 1988 (Pinto and Escobar, 1989) and concluded with the relocation of 1,117 families by 1989.

In January 2000, the Government of Guatemala began to relocate citizens of the rural town Santa Catarina Ixtahuacan due to severe problems associated with mass movements and potential landslides. Unlike other cases, this relocation took place based on scientific evidence concerning a potential disaster but not as a consequence of a disaster per se. In this process coordinated by the Executive Coordination Secretariat of the Presidency, 625 families were relocated to a new site. The process benefited from the cooperation of the Government of Germany, which contributed to the construction of the new village.

In October 2000, settlers of the rural community Santa Isabel were relocated by CONRED to a new site due to the exposure of the community to the impacts of lahars associated with the Samala River which kept destroying agricultural crops of these settlers.

In the year 2003, settlers of the rural village Texcuaco were relocated by the government to a new urban site within the city of La Gomera as Texcuaco, located on the banks of the Coyolate river, experienced floods annually.

In the year 2003, 45 of the 95 families residing in the village of El Chim, located on the foothills of Tacana volcano, were relocated due to the destruction of their houses by a massive landslide triggered by torrential rainfall.

The latest effort in terms of relocating people is taking place in Panabaj, one of the neighbourhoods of Santiago Atitlan. Located on the foothills of the Tolimán volcano, most of Panabaj was buried by a mudflow triggered by torrential rainfalls associated with hurricane Stan in October 2005.

Examples such as these serve to illustrate the fact that many villages in rural areas of Guatemala are exposed to various hazards. In most cases, relocation processes have taken place as an outcome of a disaster. However, it is interesting to note the case of Santa Catarina Ixtahaucan which was relocated on the suspicion of a potential disaster and the case of the village of El Chim which could not be relocated, despite the suspicion of a potential disaster.

2 From Risks to Relocation

In the framework proposed by the International Strategy for Disaster Reduction, ISDR, disasters are preceded by risks, which in turn are composed of hazards and vulnerabilities. The next paragraphs present definitions for these terms according to the ISDR (2008):

Hazard: 'a potentially damaging physical event, phenomenon or human activity that may cause the loss of life or injury, property damage, social and economic disruption or environmental degradation.'

Vulnerability: 'the set of conditions and processes resulting from physical, social, economic, and environmental factors, which increase the susceptibility of a community to the impact of hazards.'

Risk: 'The probability of harmful consequences, or expected losses (deaths, injuries, property, livelihoods, economic activity disrupted or environment damaged) resulting from interactions between natural or human-induced hazards and vulnerable conditions.'

Considering those communities and cities mentioned in Section 1 above, it can be concluded that all of them were at risk due to their exposure to a hazard (landslide, earthquake, lahar, etc.) and were vulnerable. As mentioned before, the invasion and subsequent conquest process, carried out by Spanish settlers, led to the establishment of communities and towns in various regions of the country which were exposed to a variety of hazards. Arriving from Spain, conquerors and settlers were not aware of the manifestation of such hazards in the new world, particularly in Central America, and thus selected sites for cities based on other factors. In a similar way, Spanish settlers introduced building techniques such as the use of adobe bricks made of dry mud and clay-tile roofs, which leads to vulnerable houses. Such a combination of exposure to seismic hazard and vulnerable infrastructure led to the disaster in 1773, where the infrastructure of the State, the Catholic Church, and the individual persons was destroyed, leading to the relocation of the city as described in the previous section. However, in the context of landslides and mudflows, even current standard infrastructure built with seismic-resistant materials may not be able to withstand the impact of such events, as manifested in the cases of Villa de Petapa, El Palmar, El Chim, and Panabaj. In these cases, relocation is usually the most appropriate solution, given the fact that it is nearly impossible to build infrastructure capable of withstanding such events. In all cases, this was the solution adopted by the government.

A contrasting example is the one related to Santa Catarina Ixtahuacan, where the relocation process took place without the manifestation of a disaster per se. In this case, citizens began to approach authorities, having witnessed particular types of damage to existing infrastructure within the community. At the technical level, the National Institute for Seismology, Vulcanology, Meteorology, and Hydrology (INSIVUMEH) was given the task of assessing the hazard related to this situation. The report prepared by the experts from this agency concluded that the town was experiencing complex mass movements associated with the local geology of the site and recommended the relocation of the community to a safer area. However, as in many other countries of the world, relocation of communities in Guatemala is a complex issue which has to be approached not only in terms of exposure to hazards but in the context of social expectations, processes, and governance at the national, provincial, municipal, and local levels. The experiences gathered in the relocation of El Chim, Panabaj, and Santa Catarina Ixtahuacan have offered a glimpse into the social issues which have to be confronted by authorities when promoting such relocation processes.

A more pressing issue in the context of small countries with increasing populations such as Guatemala is the gradual resettlement of towns and villages destroyed by natural events such as landslides, floods, and earthquakes. As mentioned earlier, towns which were destroyed centuries ago have re-emerged in recent decades and continue to expand due to a variety of factors, despite continued exposure to the hazard.

Such cases lead to the conclusion that risks associated with these events are not perceived properly. In other words, the perception concerning such risks fades away decades or centuries after a disaster, particularly when emerging issues such as poverty, vandalism, and other social problems monopolise the attention of citizens and the agenda of governments.

3 Santa Catarina Ixtahuacan, El Chim, and Panabaj

Three examples that deserve more attention in terms of the process that led to their relocation are Santa Catarina Ixtahuacan, El Chim, and Panabaj. In the case of Santa Catarina Ixtahuacan, the technical report by INSIVUMEH provided the community with a strong argument to seek the relocation of their communities. Located in the highlands of the country, in an area basically occupied by indigenous ethnic groups, settlers approached the municipal government and international NGOs for support. Unfortunately, the lack of governance in such cases led to improvisations which obstructed the process initially. A weakness which has been present in all cases associated with weak governance is the lack of legislation to deal with such relocation processes, manifested through an institutional void in terms of which agency should be responsible for management of such a process and the procedures that should be followed to reach such a decision.

An additional problem which arose in the case of Santa Catarina Ixtahuacan was the social tension which existed among ethnic groups with respect to land expected to be used for such a purpose. One potential site which was targeted by the settlers from Santa Catarina Ixtahuacan was a plateau not inhabited at that time, considered as 'safe' with respect to landslides or mudslides, and viable under the perception that such land belonged to their forefathers. Unfortunately, this land was equally claimed by citizens of the neighbouring community Nahuala, and thus, an initial occupation of this settlement led to a violent conflict which caused injuries and a few fatalities as both groups were claiming ownership. As a consequence of this conflict, the national government intervened and facilitated the new settlement on this high plateau with support of the international community.

The case of El Chim presents a similar situation regarding the lack of capacity of the government and the communities to find adequate solutions to the needs concerning relocation. While initial geological problems were identified by members of the community in October 1998, it took nearly 1 year for INSIVUMEH to conduct an inspection due to institutional weaknesses and nearly 5 years for CONRED to send a member of its risk management division to assess the risk faced by this community associated with landslides and mass movements. Both INSIVUMEH and CONRED coincided in their recommendation concerning the relocation of the community to a safer site, and the Mayor of the Municipal District of San Pedro Sacatepéquez in the Department of San Marcos was approached with the results of the assessments in order to place him in charge of dealing with the problem. In this case, problems faced by the Mayor can be summarised as follows:

- The Municipal Government has no access to public land, as most of the land is owned by individual citizens or by the private sector.
- There is a lack of policies concerning how to deal with such aspects at this municipal level. In most cases, relocations have been managed by the government at the national level.
- CONRED has a national fund established to cope with disasters, but this may not be used to deal with risks.

In this case, the Mayor needed to target social funds such as those provided by the National Social Fund or the National Fund for Peace. Unfortunately, the disaster took place before the Social Funds could provide the required funding for such a costly process.

Panabaj reflects the complexities of implementing a risk management process in the middle of a reconstruction process, where government agencies in charge of the reconstruction process (Social Funds) wish to react fast but do not necessarily take into consideration the most adequate solutions from the point of view of risk management. Panabaj represents an example of a situation where the assessment of risk conducted at the request of CONRED took longer than the time government agencies were willing to wait before proceeding with the reconstruction process. In this particular case, one of the Social Funds of the Presidency began to rebuild houses in an area that was later determined to be highly exposed to future events of this kind. The hazard assessment, conducted by an engineering firm with experience in geology and hydrogeology, took over 8 months to be completed due to the bureaucracy related to purchases and acquisition procedures. The latter were established by the government to deal with different issues, such as corruption. In this case, issues that were discussed were

- The need to maintain an image of good governance by the ruling parties, in terms of beginning reconstruction processes as quickly as possible, to avoid criticism by the media for not conducting a reconstruction process in a timely fashion.
- The need to conduct hazard assessments during the reconstruction process in order to determine areas where it may be safe to carry out reconstruction processes and areas which should be considered as unsafe.
- The lack of experience in rapidly expediting hazard assessments of this kind to provide useful information to be employed during the reconstruction process.

As it can be seen, the quick response by the government is desired by government officers who wish to keep their good image or to avoid the deterioration of such an image. As expected, any delays in the reconstruction process are viewed by the media as potential defects and as reasons to criticise such government officers. This was the case in Panabaj for several months, as the reconstruction process was literally stopped to allow geologists to carry out the hazard assessment.

4 Governance in the Context of Forced Displacements in the Case of Disasters

As several examples have shown Guatemala has undergone disasters that have forced the government to relocate entire communities, displacing citizens permanently to new sites; or disasters where such a relocation process has not been needed. Relocations can be viewed as positive from the point of view that such disasters will be avoided by minimising the exposure of the new communities to the hazards. However, in the case of disasters which have taken place centuries ago, there is evidence that people may at times reconstruct existing levels of risk by settling again in places which were destroyed centuries ago. Figure 12.1 captures these processes.

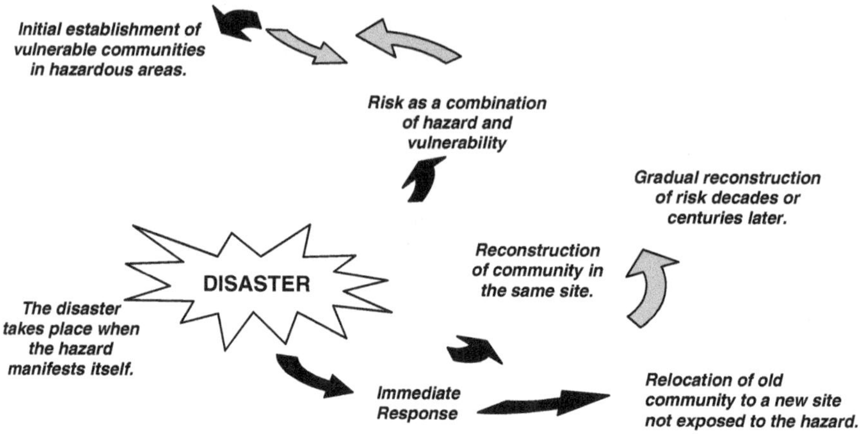

Fig. 12.1 Typical processes following a disaster in Guatemala: in some cases communities are relocated in a low hazard area, so that risks are not re-established during the reconstruction process. However, in other cases, risks have been recreated centuries later (source: authors)

In all cases where these relocation processes have taken place, the support of the national government has been a determinant, as a considerable amount of resources is required to carry out such processes. The experience in the village of El Chim is an interesting example where the community was ready to relocate to a different site, but decisions at the national level were not taken to support such local efforts. In parallel, the example of Santa Catarina Ixtahuacan suggests that the relocation process must take into consideration claims on lands which may be used for such relocation processes. Efforts taking place just at a local level in terms of the relocation of a community may be jeopardised by conflicts associated with such territorial claims. While centuries ago, land may have been available, in recent times, it may be more difficult to identify potential lands for such purposes.

A third conclusion which has been identified when analysing the way in which recent disasters have been handled is the fact that while such processes may take place more frequently than anticipated, the government has not really carried out a systematisation process in order to design both policies and rules regarding how to deal with such situations. As a consequence, improvisations are usually made, leading either to delays in the process or to inefficient use of resources.

A fourth conclusion which is drawn from the experience in the relocation process in El Palmar is related to expectations which the affected people may have concerning such a process that may not materialise as expected. In the case of El Palmar, the relocation process meant that the Government would exchange houses in the old site with houses in the new site. However, in some cases, there were two or even three families living in one house in the old El Palmar, each of which had the expectation of receiving its own house, rather than having to share a new house in the new El Palmar.

One critical aspect which has received little attention in such cases is the ownership of land that is vacated. During the relocation process, families are provided titles with respect to the new houses which they are receiving, but in many cases they have not been asked to relinquish their claims on properties which were destroyed or affected by the disaster that led to the relocation. A case in point is Santa Catarina Ixtahuacan, where the former village is beginning to be occupied through two parallel processes:

- New residents migrating from other areas, who purchase old property rights from families that benefited from leaving their new houses in the new village.
- Old residents or relatives, who may wish to take advantage of the fact that the old site did not experience yet the foreseen disaster and who may additionally wish to continue benefiting from the new land. In essence, this means that families who experience a disaster may end up with claims to properties both in the old and in the new sites.

Such processes have also been taking place in other regions of Central America which have been devastated by earthquakes and are highlighting a void with respect to how to manage such issues.

5 Policy-Relevant Recommendations

Taking into consideration critical aspects mentioned in previous sections related to relocation processes, the following policy-relevant recommendations stand out:

- Appropriate policies and rules are required to delegate the responsibility concerning such relocation processes to either a single government agency or a group of agencies which need to cooperate in order to complete the relocation processes adequately. Such policies should dictate how governments at different levels, and particularly with different autonomies, should work together with local communities and civil society in order to conduct such relocation processes efficiently and in a timely way.
- Operational procedures within the Social Funds operated by the Presidency of the Republic should be modified in a way that risk management is incorporated within the funding practices, so that relocation processes can be considered and implemented when such a need arises.

Awareness campaigns should be targeted to make the population aware of risks in particular locations; as well as to keep the social memory of historic events, so that such events do not repeat themselves in terms of catastrophic consequences.

The first recommendation should address the following questions:

- Which agency should be responsible for starting such processes?
- When a group of concerned citizens is willing to promote such a process, which agency should they target so that their request can be conveyed to the proper agency in charge of such processes?
- What should be the role of different civil society organisations, such as the Church?

The second recommendation targets those agencies of the national government that are entrusted with financial resources to carry out projects to promote sustainable development. In this context, it is equally important to assess the role of such funding programmes in promoting the use of information concerning risks to assess the need to relocate communities or not.

The last recommendation is aimed at ensuring that the population is aware of the risks which lead to particular disasters so that such conditions of risk are not recreated as a result of lack of awareness.

6 Conclusions

Due to its geographical location and topography, Guatemala is exposed to a variety of hazards which have triggered disasters leading to temporary and permanent displacements of families and communities. Historical and recent relocation processes resulting from such disasters have been handled by the national government, but the lack of systematisation and proper institutionalisation of such processes continues to inhibit the government from conducting such processes in an efficient and timely way.

As stated by the *International Organisation for Migration* (IOM, 2001), migration and relocation processes are constrained by factors such as their dimension, social organisation, political aspects, and local development. While forced displacements have been triggered by disasters, voluntary migrations within Guatemala are rooted in internal factors such as poverty and the inability of citizens to meet their basic needs; and on external factors such as the availability of opportunities to migrate in search of better options either in the capital city or in the United States.

An analysis of this complex problem of relocations has led to the identification of policy-relevant recommendations which should be taken as inputs by government agencies in Guatemala, as well as in countries which face similar problems, to conduct such processes in a more efficient way in the future, particularly as the degradation of the environment will surely contribute more and more to phenomena such as landslides in rural and urban areas that are at the root of such forced displacements.

References

CONRED. (1997): Declaratoria de Alta Riesgo del Antiguo Palmar, Quetzaltenango.

Government of Guatemala. (1987): Acuerdo Gubernativo Número 1–87. Published in Diario Oficial, Guatemala 1987.

Government of Guatemala. (2001): Decreto Número 5-2005. Published in Diario Oficial, Guatemala, September 2001.

IOM. (2001): Desastres y Migraciones en Guatemala. Cuaderno de Trabajo sobre Migración No. 3. Available from http://www.oim.org.gt/Cuaderno%20de%20Trabajo%20No.%2003.pdf

ISDR. (2008): Terminology: Basic terms of disaster risk reduction. *International Strategy for Disaster Reduction*, The glossary containing definitions of terms. Available from http://www.unisdr.org/eng/library/lib-terminology-eng%20home.htm

Juarros, D. (1936): Compendio de la Historia de la Ciudad de Guatemala. Volume 1, Tipografia Nacional de Guatemala, Third Edition, 1936.

OFDA-CRED. (2007): "EM-DAT: The OFDA/CRED international disaster database" Université Catholique de Louvain, Brussels, Belgium. Data from historical disasters in Guatemala corresponding to the period 1902–2007. Available from www.em-dat.net

Pinto, J.C., and M.T. Escobar (1989): El Palmar: ¿solo un desastre natural? Bulletin No. 3, CEUR, USAC, February 1989. Available from http://www.usac.edu.gt/~usacceur/pdf/Boletin/Boletin_CEUR_03.pdf.

Prensa Libre. (2001): "Hambruna, decretan Estado de Calamidad". Prensa Libre Newspaper edition corresponding to September 4, 2001. Available from http://www.prensalibre.com.gt/pl/2001/septiembre/04/11094.html

Part V
EACH-FOR Case Studies and Scenarios

In this section of the book, there are seven chapters written by researchers in the Environmental Change and Forced Migration Scenarios (EACH-FOR) project (www.each-for.eu), which was supported by the European Commission in the Sixth Framework Programme and ran from January 2007 until March 2009.

The EACH-FOR project had two central aims: (1) to explore and describe the causes of forced migration in relation to environmental change; and (2) provide plausible future scenarios of environmentally induced forced migration. These specific objectives were achieved by the systematic overview and analysis of the relevant natural and human-made environment degradation processes, as well as the socio-economic and demographic contexts, in the regions studied in the project; fieldwork or desk studies in 23 case study areas; and subsequent scenario development for 6 of the case study areas.

Six of the 23 case studies are briefly described here: Mexico, Senegal, Egypt, Bangladesh, Turkey, and Kyrgyzstan. This provides an example from each of the regions outside of Europe and Russia that were covered by the EACH-FOR research. It also provides various examples of environmental degradation (floods, droughts, land degradation, sea-level rise, tropical cyclones, nuclear waste, and landslides) that affect human livelihoods. The chapter on the Turkey case study shows not only the effects of overuse of ground water but also the particular case of forced displacement due to a development project.

The final chapter in this section of the book illustrates the approach taken in the EACH-FOR project to develop scenarios. Noting that scenarios are *not* predictions, the chapter describes a methodology for deriving local-level scenarios that can be used to discuss what migration patterns could evolve depending on different development pathways in the future.

Taken together, these chapters illustrate some of the main conclusions of the EACH-FOR project:

- Migration decisions are complex reflecting the interconnectedness of environmental factors with economic, social, and political factors;
- Climate change is not the only potential trigger for migration;
- Migration is a traditional coping mechanism but in some areas these traditional patterns have changed in recent decades;

- Migration occurs when livelihoods cannot be maintained, especially when agriculture or herding is severely affected by environmental degradation or extreme events;
- People who want to leave their villages/regions/country can only do so if they have the necessary financial means and access to networks that support migration;
- In some cases there are no 'winners' – both those who migrate and those who stay are losers.

The study of forced displacement as a result of dam construction provides valuable lessons regarding the resettlement process, in particular the need for participatory processes with significant support and information for those being resettled.

Chapter 13
Environmental Factors in Mexican Migration: The Cases of Chiapas and Tlaxcala

Stefan Alscher

1 Introduction

Even though Mexican migration is one of the most cited cases in migration research, the impact of environmental degradation or natural disasters on migration flows originating in Mexico is still a field under exploration. This chapter, based on fieldwork carried out in the frame of the EACH-FOR Project, gives an overview of selected problems of environmental degradation, of migration flows within and from Mexico, as well as of linkages between both phenomena.[1] Two cases have been selected in this study, illustrating the linkages between environmental problems and migration. In Western Tlaxcala, located in central Mexico, the consequences of desertification, land degradation, and changing rainfall patterns on the rural economy and on migration are the main focus. In the case of the south-eastern state of Chiapas, the poorest state within the Mexican federation, the impact of tropical storms on migration flows is the focus.

2 Synthesis of Context

As one of the most cited cases in migration research, nearly every single aspect of Mexican migration has been studied in the last few decades. Nevertheless, the research on environmental migration or – more precisely – the impact of environmental degradation, natural disasters, and climate change on internal and international migration flows originating in Mexico is still a field under exploration. Some research has been done already, even though not all studies were implicitly on the link between environmental change and migration.

S. Alscher (✉)
Center on Migration, Citizenship and Development (COMCAD), University of Bielefeld, Bielefeld, Germany
e-mail: stalscher@migra.org

[1] A more detailed version, including a list of interview partners in Mexico, is available on the EACH-FOR website (www.each-for.eu).

T. Afifi, J. Jäger (eds.), *Environment, Forced Migration and Social Vulnerability*, DOI 10.1007/978-3-642-12416-7_13,

Already in the late 1970s, Medellín (1978) estimated that around 600,000 Mexicans were abandoning the countryside each year, because of the inability to subsist on agriculture as formerly workable lands became unproductive due to desertification. A report by Campbell and Berry (2003) mentioned a range of 700,000–900,000 Mexicans who were leaving their lands every year because of land degradation and desertification. Also the US Commission on Immigration Reform assumed a linkage between environmental degradation and migration and therefore requested a study on unsustainable land and water use as a potential factor driving Mexican migration to the United States. This study, carried out by Leighton-Schwartz and Notini (1994), came to the conclusion that further degradation of Mexican farmland would increase internal and international migration flows. The authors presented some evidence that a part of the contemporary migration to the United States was due to the impact on farmlands of increasing desertification. Similar findings were made by a follow-up study of the National Heritage Institute on desertification and migration in Mexico, which came to the conclusion that 'there is a strong correlation between environmental stress, poverty and population pressure, which can lead to migration' (NHI, 1997). De Janvry et al. (1997: 17) argued that environmental stress variables, such as deforestation and population pressure on limited fertile lands, create an incentive to migrate. Political measures against deforestation and land degradation would therefore be an effective measure to reduce incentives for migration. Escobar et al. (2006) analysed the impacts of hurricane Mitch in selected regions of Chiapas, taking into account the fragile economic context of Chiapas and the crisis in Mexican agriculture. Together with these socio-economic factors and insufficient state support, the devastating effects of Mitch became an additional reason for emigration. Also, Villafuerte and García (2006) argue that the effects of hurricane Mitch – and probably also of Stan – were an additional factor for emigration, besides the rural crisis since the 1980s and the political and military conflict since the Zapatista rebellion in 1994. According to Saldaña-Zorrilla (2007: 27 f.), losses from natural disasters exceed rural coping capacities. As state support for the countryside is insufficient, rural workers emigrate to urban centres or towards the United States.

Taking the Gross Domestic Product (GDP) as an indicator, Mexico is the 2nd-largest economy in Latin America (after Brazil) and the 13th-largest economy worldwide (World Bank, 2008). According to data from the Economic Commission for Latin America and the Caribbean (ECLAC), Mexican GDP was 839.5 billion USD in 2006 (at current market prices; ECLAC, 2008: 89). Even though Mexican policymakers always tend to underline growth rates and other overall economic figures, the economic growth did not lead to a significant improvement of life conditions for huge parts of the Mexican population. After three decades of high economic growth (annual growth rates of 6% or higher), characterised by the policy paradigm of import substitution industrialisation, the Mexican economy fell into a crisis in the 1970s, culminating in the declaration of insolvency and a stop in the payment of foreign debts in 1982.

Since the presidency of Miguel de la Madrid (PRI[2] 1982–1988), a new political elite prepared a policy change towards a model of neoliberal economic policies. The entrance of Mexico into the General Agreement on Tariffs and Trade (GATT) in 1986 and the negotiation process of the North American Free Trade Agreement (NAFTA) under President Carlos Salinas (PRI, 1988–1994) in the early 1990s were central pieces of the new economic policy. NAFTA came into force on 1 January 1994 – on just the same day when the Zapatista (EZLN) insurgency started in the south-eastern state of Chiapas. The concurrence of these two events can be deemed as symptomatic of the socio-economic conditions in Mexico: on the one hand, a paradigm of free-market policies and increasing participation in the global market occurred, while on the other hand, poverty and inequality in broad sectors of the Mexican society prevailed or even increased.

Since the NAFTA came into effect, the social and economic division lines within Mexico became even more pronounced. While northern Mexico and greater urban centres indeed experienced some market-induced economic growth, the rural areas – above all in southern and south-eastern Mexico – were not able to compete with the US-American agro-business. One of the clearest symbols of the structural inequality within NAFTA is the increase in the export of corn (Mexican: *maíz*) from the United States to Mexico. Corn, a major component of the daily Mexican diet for centuries, is the main source of income for about 40% of those working in agriculture (Raghavan, 2000). The planned transition period for the liberalisation of corn imports was planned for 15 years, but in reality it was compressed to about 30 months, which meant an extremely high pressure for small-scale farmers in Mexico (Nadal, 2000: 31). Since the early 1980s, Mexican agriculture experienced a profound crisis. Market opening and deregulation of the economy were accompanied by the elimination of state subsidies for Mexican farmers (Villafuerte and García, 2006: 104), affecting mainly small producers. As a consequence, poverty – as well as internal and international migration – increased. According to the World Bank, 57% of the rural population lives in conditions of poverty and 28% in extreme poverty (World Bank, 2004). In its country report on Mexico, the Social Watch points out that poverty reduction measures were only able to re-establish the poverty levels that prevailed prior to the economic crisis in the mid-1990s (Social Watch, 2008).

2.1 Overview of Environmental Problems

The central problems of environmental degradation in Mexico are mainly interlinked with the degrading quality of soil and insufficient availability of water. Deforestation, extensive agricultural production, salinisation, and other

[2] PRI: *Partido Revolucionario Institucional*/Institutional Revolutionary Party, governing party from its founding in 1929 until the elections in 2000 (former names: National Revolutionary Party, Party of the Mexican Revolution).

factors led to soil degradation and desertification. More than 85% of Mexico is affected by desertification, above all in the fragile arid and semi-arid ecosystems of northern and north-western Mexico – where more than 60% of the land is considered to be in a total or accelerated state of erosion. However, former fertile lands in the mountainous regions of central Mexico and in the tropical south and southeast of Mexico are also increasingly exposed to soil degradation, which is mainly due to the cutting down of vast forest areas. This contributed to changes in regional rainfall patterns, producing a vicious circle of decreasing precipitation and increasing desertification. The pollution of soils, water, and air is another striking problem, above all in and around urban centres.

Mexican climatic zones span from the hot, dry conditions of the north-western Sonoran desert (annual average rainfall less than 100 mm) to the wet, tropical conditions that characterise the forest regions of southern and south-eastern Mexico, above all Chiapas and the Gulf coast states of Tabasco and Veracruz (average annual rainfall can reach 2,000 mm or even more). According to data of Mexico's National Meteorological Service (SMN), drought periods are becoming longer and more extreme in several regions of the country. This trend not only affects the arid and semi-arid regions of the North and Northwest but also regions in the humid South. Another phenomenon is the shifting of rainfall periods, as the example in the field study region of Tlaxcala shows. While the rainfall period used to start in March in former times, nowadays the rainfall starts about a month later. In the case of the traditional product of corn, this means that the harvest period shifted from June to late July or even August. In other regions, the trend seems to be just the other way round: due to the higher intensity of tropical storms, the amount of heavy rains has increased in some regions of south-eastern states like Chiapas and Tabasco.

As stated above, deforestation is another central problem in Mexico. According to the Ministry for the Environment and Natural Resources (SEMARNAT), Mexico lost around 1.1 million hectares of forest per year from 1993 to 2000. Therefore, Mexico is on the second place worldwide in deforestation, just behind Brazil. If this trend continues, Mexico will lose all of its tropical forests – now covering around 30 million hectares – in the next 52 years (WRM, 2002). Other types of forests, now covering around 33 million hectares, would disappear in about 122 years. Main causes for the disappearance of the forests are burning, logging, and fuel – and above all the increased use of land for agriculture and cattle, which is also partially due to the changes in the economic model (market liberalisation, free-trade agreements, etc.). According to Wise and Gallagher (2002), the increased pressure from US exports is not only a cause for emigration but also threatens the biodiversity in Mexico, as the growing competition increased the use of chemical-intensive substances. Furthermore, corn production has been expanded to drier states, resulting in an unsustainable use of water for irrigation. Instead of an upward harmonisation of environmental standards – which in fact is part of a NAFTA side agreement – environmental degradation has accelerated since the implementation of free-trade policies in the mid-1980s. Between 1985 and 1999, rural

Table 13.1 Summary of top five natural disasters in Mexico (1900–2008)

Event	No. of events	Killed	Affected	Damage in tsd. US$
Storm	72	5,052	6,451,315	15,310,810
Earthquake	27	10,677	2,556,577	4,691,000
Flood	49	4,103	3,167,344	4,533,400
Drought	6		65,000	1,610,000
Volcano	10	1,120	161,908	117,000

Source: EM-DAT: The OFDA/CRED International Disaster Database, www.emdat.be – Université catholique de Louvain – Brussels; accessed on 6 May 2008.

soil erosion grew by 89%, solid waste by 108%, and air pollution by 97%. The economic costs of environmental degradation have been estimated by the Mexican government to be around 10% of annual GDP (Wise and Gallagher, 2002).

Besides the processes of slow-onset environmental degradation, many regions of Mexico are frequently hit by natural hazards, such as hurricanes, earthquakes, and floods. Hurricanes and other tropical storms are the most significant type of disaster, in terms of economic damage and the number of people affected (see Table 13.1). This is due to the geographic location of Mexico, being on the path of hurricanes coming from both sides, the Pacific Ocean and the Gulf of Mexico as well as the Caribbean Sea. Since the second half of the 1990s, the frequency of hurricanes and other tropical storms has increased, especially in the south-eastern states of Yucatan Peninsula and Chiapas. Recent hurricanes which caused massive destruction include Mitch (in 1998), Stan and Wilma (both in 2005). In the case of south-eastern Mexico, hurricanes and other tropical storms frequently led to floods and landslides, also fostered by deforestation and soil erosion in hillside regions.

The highest number of people killed by natural hazards has been registered with earthquakes, which is mainly due to the 1985 Mexico City earthquake (19 September 1985), when – according to governmental statistics – over 9,000 people were killed, 30,000 injured, and around 100,000 left homeless. The Seismological Service at the National Autonomous University (UNAM) estimates the number of killed people to be as high as around 40,000 (SSN, 2007). An analysis of population statistics shows that the 1985 earthquake also resulted in a lower rural–urban immigration from the Mexican countryside to Mexico City and some out-migration from Mexico City. While in 1980, the total population of the Federal District (= Mexico City) reached more than 8.8 million people, this decreased to 8.2 million in 1990. This was the first time in modern history that a decrease of the total population of Mexico City had been registered.

Floods also caused high damages in some Mexican regions. The most recent and internationally well-known case is the 2007 Tabasco flood, caused by the tropical storm Noel in the Caribbean Sea. After several days of heavy rains in Tabasco and the neighbouring state of Chiapas, overflowing river arms of the Grijalva led to floods in some municipalities. On 29 October 2007, the Federal

Commission of Electricity (CFE) announced a partial emptying of the Peñitas Dam in Chiapas and of other dams in northern Chiapas the day after, adding more than 2,000 m^3 water per second to the Grijalva river (ADN Mundo, 2007). Together with continuing heavy rainfall, the water masses led to the inundation of about 80% of the state of Tabasco, including the capital city Villahermosa. According to the Emergency Events Database (EM-DAT), around 1.6 million people were affected by the flooding, and an economic damage of 3 billion US$ was recorded. The heavy rainfalls and the partial opening of dams were not the only reasons for the huge floods; deforestation in Chiapas and Tabasco, the lowering of sediments through the massive exploitation of oil fields in Tabasco, and the non-fulfilment of promised public investments in the hydrological system were other important factors.

Another environmental challenge for several coastal regions of Mexico is the probability of a sea-level rise. Ortíz and Méndez (1999, 2004) show that low-lying areas of the Gulf Coast and the Caribbean are especially vulnerable to sea-level rise. Above all, in the case of the lowlands of Tabasco, such scenarios are highly critical, as vast areas (up to 40–50 km inland) would be affected. This is also the most populated of the vulnerable areas, with Villahermosa as a big city (615,000 habitants) within the area of risk.

2.2 *Overview of Migration Processes*

Emigration from Mexico is overwhelmingly directed towards the United States (other international migration flows are close to insignificant) and strongly interlinked to the respective economic, social, and political conditions in both countries. The history of Mexican migration to the United States goes back as far as to the mid-nineteenth century, when Mexico lost about 40% of its former territory after the Mexican–American War (1846–1848). Thousands of Mexican families stayed in their villages, located since then on US territory, which became the basis for the first cross-border social networks. Until the early twentieth century, Mexican emigration to the United States remained at a very low level, with Mexican migrant workers being employed in the construction of railways, mining, and agriculture in the Southwest. In the first decade of the twentieth century, the economic and agrarian modernisation that uprooted parts of the rural population not only served as a precondition for the revolution but also led to a potential of (former) peasants willing to leave their communities. The political, economic, and social instabilities during the revolutionary years (1910–1917/1920) as well as post-revolutionary conflicts led to further migration. From 1900 to 1930, around 730,000 Mexicans migrated to the United States. The growth in the plantation economy and new jobs in the industrial and service sector in the US Southwest also contributed to the increase of migration. Mexicans were perceived as an ideal agricultural workforce, mainly because of the geographical proximity, which facilitated both recruitment and – during economic downturns – deportation of migrant workers.

With the entrance of the United States into the Second World War, economic growth took off and the foreign labour force was needed more than ever. US employers sent recruiters to Mexico, a process which became formalised with the signature of the so-called bracero agreement (signed on 4 August 1942).[3] This program laid the basis for a further consolidation of dense migration networks between regions of origin in Mexico and regions of destination in the United States. In the 22 years of its validity, about 4.6 million contracts were issued, but as many bracero workers came to the United States repeatedly, the total number of Mexican who benefited workers is estimated to be between 1 and 2 million (Martin and Midgley, 1994: 19). When the US economy entered a recession after the Korean War, several hundreds of thousands of Mexicans were deported or forced to leave in the so-called Operation Wetback. In the following decades, Mexican emigration towards the North experienced a continuous growth, partially due to political reforms in the United States (1965 amendment of the Immigration and Nationality Act of 1952 and legalisation programme under the Immigration Reform and Control Act of 1986). But also the economic and demographic context in Mexico contributed to further migration flows, such as the crisis of the economic model of import substitution industrialisation and the steady population growth. This led to a saturation of labour markets and increased the attractiveness of migrating towards the North.

The reduced possibilities to immigrate legally to the United States resulted in a change of migration patterns, fostering undocumented migration. This increase is clearly reflected in the detention statistics of the US Border Patrol, even though these statistics only reflect a small part of reality. The number of detained undocumented migrants was around 110,000 in 1965, more than doubled to 284,000 in 1969, surpassed 1 million in 1979 and reached a preliminary peak in 1986 with nearly 1.8 million detentions (Calavita, 1992: 217). The overwhelming majority of undocumented migrants detained by the Border Patrol was (and still is) of Mexican nationality. The steady increase of undocumented migration led to the implementation of border protection programmes since the mid-1990s, including the construction of physical barriers on the borderline. It is interesting to note that only a few months after the construction of a first border fence between San Diego/USA and Tijuana/Mexico (since March 1993), the North American Free Trade Agreement (NAFTA) entered into force (1 January 1994). While NAFTA promotes a liberalisation of the free movement of capital, services, and goods, border fences and walls have been erected in order to contain – or at least shift – flows of unwanted migration of labour force, which also led to an increase of border deaths (Andreas, 2000; Cornelius, 2001).

---

[3] Official name: Mexican Farm Worker Program, 1942–1964.

While Mexican migration from the 1940s to 1960s was mainly concentrated on some regions in Northern and Central Western Mexico, i.e. those regions that participated strongly in the bracero programme, nowadays Mexican migrants come from nearly every corner of the country. The diversification of Mexican migration is mainly due to economic downturns in the 1970s (crisis of ISI model), 1980s (crisis of Mexican agriculture), and 1990s (peso-shock 1994), which affected large parts of the population. Together with the already existing networks and the negative impacts of free trade for rural areas, migration became a seemingly attractive option for large parts of the population. Nearly all Mexican families have some relatives living in the United States. In many municipalities, the share of those households receiving remittances is between 20 and 25%, in some even 40% or higher.

Even though reception of remittances may be a useful indicator for the extent of migration, it is difficult to break down the intensity of migration processes to the sub-national level. Official statistics from the National Population Council (CONAPO) seem to lag behind the social reality. States like Chiapas and Tlaxcala still appear as regions with low rates of emigration, although all interview partners painted a different picture. In the case of Tlaxcala, only the municipality of Hueyotlipan is labelled with a 'high grade of migratory intensity'; in the case of Chiapas, all municipalities are marked as 'low' or 'very low' regarding the intensity of migration. The picture becomes different when comparing data from the 2000 general census with the 2005 population count. A clear example is the municipality of Motozintla, one of the municipalities which was heavily affected by hurricanes Mitch (1998) and Stan (2005). From 2000 to 2005, the total population of Motozintla decreased by 3%, the male population even by 5%. This drop in total population can only be explained by emigration, as other municipalities in the same state that were not affected by the hurricanes, have experienced a regular population growth. Other municipalities[4] with a male population decrease of more than 5% are mainly located in the Soconusco and Sierra region of Chiapas, where hurricane Stan had the strongest impact (INEGI, 2000, 2005). Breaking down the figures to an even smaller level, i.e. communities within the municipality, the population change between 2000 and 2005 becomes even more obvious. In the three municipalities which were visited during the fieldwork (Huixtla, Motozintla, and Tapachula), dozens of communities had a decrease of population in the period 2000–2005, while the same communities still experienced a population increase in the period 1995–2005. In several cases, the loss of population was higher than 50%, while the average population growth in the state of Chiapas has been around 10% in the 5 years from 2000 to 2005 (nationwide 5.9%). Some communities were even not mentioned anymore in the 2005 population count, which may be not only due to purely statistical reasons (communities with less than 3 houses

[4] Acatepahua, Amatenango de la Frontera, Bella Vista, Chiapilla, Escuintla, Huehuetán, Mapastepec, Ostuacán, and Unión Juárez.

are not disaggregated), but also due to problems of accessibility after the passage of Stan, or in the worst case due to the fact that the community just disappeared during the storm and flooding. Mapping the communities with a population decrease, a clear parallel with the most affected areas becomes obvious.[5]

In the case of Tlaxcala, the analysis of population data from the Mexican statistical institute INEGI[6] also shows some differences from the data of the National Population Council (CONAPO). Only Hueyotlipan has been classified by CONAPO as a municipality with a high intensity of emigration. Looking on the level of communities, it is above all San Simeón Xipetzinco which experienced a high decrease of population between 2000 and 2005 (−18.4%). Several other communities show high differences between changes of male and female population; with a decrease of the male population and an increase of the female population, which is an indicator for emigration. While the decrease of the male population in the whole municipality of Hueyotlipan has been only −1.35% between 2000 and 2005, this figure is much higher in other municipalities, which according to CONAPO only shows low or very low migration intensity. The relatively low decrease registered by the population count in Hueyotlipan may be attributed to the predominantly temporary character of migration from this community (labour agreements). Migrants are still residing in their community of origin for several months per year and therefore are also counted as residents. Even though the decrease of rainfall as well as the increase in out-migration started in the 1990s in the case of Tlaxcala, a clear linkage cannot be drawn on the basis of statistics, as the data sets are too sparse for that purpose (no annual migration data on community or municipal level).

3 Methods and Selection of Regions

The selection of both cases, Chiapas and Tlaxcala, was based on a review of literature and on talks with academics in Mexico. Despite a bad performance regarding socio-economic indicators (lowest GDP per capita and lowest Human Development Index within Mexico), Chiapas is not a 'traditional' region of emigration. Located on the route of many tropical storms, the region is frequently suffering from the consequences. The impacts of tropical storms have even become worse in recent decades due to deforestation and soil degradation, which led to a higher vulnerability in the case of heavy rainfall, especially to a higher frequency and stronger impacts of floods. The other selected

[5] It is important to mention that, in the case of Chiapas, the realisation of the 2005 count was prolonged until 15 November 2005, due to the difficulties in reaching remote communities after the passage of hurricane Stan in early October 2005. As the time period between the destruction caused by the storm and the realisation of the survey is very short (6 weeks), it is questionable, to which extent the impact of Stan may be reflected in the 2005 data.

[6] *Instituto Nacional de Estadística y Geografía.*

case, western Tlaxcala in central Mexico, is ranked 24th among the 32 Mexican states regarding human development (PNUD, 2007: 23). Agricultural production in Tlaxcala depends heavily on rainfall patterns which have changed in the course of the last two decades. According to researchers of the National Autonomous University (UNAM), Tlaxcala has become the nationwide number one in processes of desertification, mainly due to the intensive use of soil. The state is also considered to be very vulnerable to the effects of climate change (Hérnandez et al., 2003: 36). Even though Tlaxcala is not a traditional emigration state, internal and international migration flows from this area have increased since the 1980s.

During the fieldwork, expert interviews with academics gave an overview and facilitated further contacts (snowball method). The next steps consisted of contacting key persons (e.g. local researchers, NGO activists) who were then interviewed during the research in the selected regions and helped in the identification of those communities that are particularly affected by environmental degradation. In most visited communities, the commissioners of the communal 'ejido' lands (*comisariado ejidal*) were of central importance, since they shared their knowledge of social, economic, and environmental problems in their communities and helped in identifying and contacting interviewees. Semi-structured interviews have been realised with most contacted commissioners. The final step was the application of the standardised EACH-FOR questionnaires and qualitative interviews with non-migrants.

4 Fieldwork Findings and Analysis

In the case of the analysed areas in southern Chiapas (communities in the coastal municipalities of Huixtla and Tuzantán, in the mountain municipality of Motozintla and in the urban municipality of Tapachula), the impact of both slow-onset environmental degradation as well as natural hazards on migration processes can be clearly seen. According to several interviews, people were already planning to leave their communities before hurricane Mitch (1998) and hurricane Stan (2005), respectively, passed through their region, but the severe destruction caused by these hurricanes accelerated the decision to migrate, as the economic basis of many families was totally destroyed after the passage of the hurricanes.[7]

Asking the villagers about migration, most respondents underline that they have no other place to go. Furthermore, the age of the interviewees plays an important role in the decision to migrate. In small communities like Toliman or Villahermosa (Motozintla), most young people already left in recent years. The older ones prefer to stay in their villages and continue with their activities. Those with relatives abroad receive remittances, which have become an

[7] Interview with Hugo Ángeles, ECOSUR Tapachula.

important source of income in the affected areas. Asked about what would have to happen in order that they decide to migrate, the typical answer was that their house and/or lands would have to disappear. Regarding governmental support, villagers in all three areas (mountains, coastal plains, and suburban shantytowns) criticised the state institutions. Even though some interviewees pointed out that they received some financial or material help for reconstruction, most of them stated that governmental help came far too late or even not at all. Another problem mentioned by several villagers was the assignation of houses for environmentally displaced persons. Governmental authorities built houses in the municipality of Tuzantán for those who lost their homes during hurricane Stan. Some villagers complained that the assignation of houses was not regulated in a fair manner, but most complained about the conditions of these houses (missing electricity, insufficient sewage system, fragile construction). For these reasons, several interviewees stated that they would rather not accept the houses but live with other family members, closer to their community of origin.

In several communities, deforestation and soil erosion were an aggravating factor when hurricanes or other tropical storms passed through. The mountain region in southern Chiapas used to be covered by dense forests, which also served as a protection against extreme weather events. According to experts, about 76% of the forest coverage in Chiapas is degraded (Greenpeace, 2005). In those areas where deforestation took place, devastation after the tropical storms is more extreme than in those areas with intact forests. A major problem of deforested hills is the high risk of massive landslides; in some cases, entire villages were buried under the masses of mud. Moreover, landslides are a substantial threat for local agriculture, as plantations have been buried by mud after heavy rainfalls.

The findings from the qualitative interviews with villagers and local agents are also backed by the questionnaires. A total of 39 questionnaires have been applied in different communities in the municipalities of Huixtla, Tuzantán, Motozintla, and Tapachula, 20 of which were 'migrant questionnaires' and 19 'non-migrant questionnaires'. The majority of respondents acknowledge that environmental problems have affected their livelihood and also see a relation between environmental degradation and migration; 19 out of 20 migrants stated that environmental problems affected their decision to move, mainly because of sudden natural disasters. Among the non-migrants, sudden natural disasters and unreliable harvest were identified as the main reasons for migration. Nevertheless, other factors also played an important role, such as the crisis of Mexican agriculture since the 1980s and the negative impacts of free-trade policies for rural Mexico. The decline of prices for agricultural products, especially in the case of coffee, together with the elimination of state subsidies and growing competition from abroad, led to a sharp decrease of family income in rural areas. This leads to the diversification of income sources, mainly by sending younger family members away, either to other regions within Mexico or to the United

States. Data from the EMIF-survey[8] show that migrants from Chiapas only represented around 1% or less of those captured by the survey ('migrants coming from the south'), while they already represented 4.6% of the total in 2002/2003. The increase becomes even clearer when considering those migrants who are looking to migrate to the United States: less than 1% until 2000 up to 7.2% in 2002/2003 (COLEF/CONAPO, 2006).

In the case of western Tlaxcala, the relationship between environmental degradation and migration is not as obvious as in the case of Chiapas. Most interviewees, migrants as well as non-migrants, emphasised that their lands still have a 'good quality'. Low prices for their agricultural products, division of lands through heritage, and lack of interest of the younger generation for agriculture were seen as the main problems. But in the course of the interviewing process, the majority of interviewees complained about the shifting of rainfall periods, which results in a decrease of harvests and therefore in lower earnings. Some also admitted that techniques of modern agriculture led to soil degradation. Leaders of peasant organisations expressed concern about the degrading quality of soils in the region. When asked about the emigration of young villagers, NGO leaders coincided with the interviewed peasants that the younger generation is looking for other jobs than working in agriculture, but stressed that this is linked with low earnings in agricultural production as well, resulting partially from the shifting of rainfall periods. Therefore, the linkage of environmental degradation and migration can be characterised as indirect in this case. Summarising the statements, emigration is a strategy for income diversification. This diversification is gaining more and more importance, as the villagers are confronting a set of adverse conditions, such as low market prices for their products, elimination of state support, and environmental factors, such as soil erosion and change in rainfall patterns. It is interesting to note that one interviewee mentioned inundated farmlands. This may sound paradoxical at first sight but is in fact a real problem for the villagers; if heavy rains come after a long period of drought or just after sowing the first plants, the impact for the farmers becomes hazardous, as the fertile layer of the soil is washed away. Finally, the use of modern agriculture also aggravated the problems in the two selected communities in western Tlaxcala. The *Ejido*-commissioners of Benito Juárez and Hueyotlipan coincided in their view that the use of heavy machines and chemical fertilisers led to a higher risk of soil degradation. On the other hand, these machines and fertilisers are perceived as a requirement in order to be able to compete on the market, which creates a vicious circle for the farmers in several regions of Mexico.

---

[8] *Encuesta sobre* Migración *en la Frontera Norte de México*/Survey on migration at the Northern Border of Mexico.

5 Conclusions and Future Research

The fieldwork carried out in two different regions of Mexico not only shows some clear indicators for environment-induced migration, but also makes clear that it is difficult to define the role of environment as a push factor within the broad spectrum of other, above all socio-economic, conditions. The selected cases show clear differences in the importance of environmental factors regarding the decision to migrate. In the case of Chiapas, the hazardous impacts of hurricanes had a 'trigger' effect for internal and international migration. In Tlaxcala, soil erosion and changing rainfall patterns serve as an additional push factor for emigration. In both cases, a set of other factors is of crucial importance for explaining migration, especially the crisis of Mexican agriculture, growing international competition, missing state support, and low prices for agricultural products. Nevertheless, environmental conditions also play a central role, as they are the basis for agricultural production. Both states are recent emigration states within Mexico. Mainly young villagers see emigration as a viable option. Emigration serves as a strategy of income diversification; remittances are mainly used for the basic needs of the remaining family members.

Processes of slow-onset degradation as deforestation, soil erosion, and desertification affect large parts of the country (Martínez and Fernández, 2004). Many regions are also frequently hit by natural disasters, such as hurricanes, earthquakes, and floods. For some coastal regions, the probable sea-level rise is a serious challenge. Even though the Mexican government has an organised structure for disaster prevention and response, reconstruction of affected regions varies from one region to the other, mainly due to economic interests. It is obvious that this analysis only reflects a small part of the reality. Further research would be necessary in order to get more evidence on the linkages between environmental change and migration. This research should focus on other regions in Mexico, for example, those which already reached a high level of desertification or those with a longer history of emigration. Another interesting point would be research in the state of Tabasco, where large parts of the state have been inundated during the floods in 2007 and where the risk of further inundations due to erosion and sea-level rise is very high.

Acknowledgements The author wants to give his special thanks to collaborating researchers in Mexican partner institutions, above all to Hugo Ángeles, Martha Rojas, and Carmen Fernández from the El Colegio de la Frontera Sur (ECOSUR) in Tapachula, and Fernando Herrera and Oscar Calderón from the Universidad Autónoma Metropolitana – Iztapalapa (UAM-I) in Mexico City. Their support has been of central importance for the realisation of fieldwork in the selected regions. This research was carried out within the framework of the EACH-FOR Project with the financial support of the European Commission's Sixth Framework Programme (contract number 044468). The project has been realised between January 2007 and March 2009.

References

Mundo, A.D.N. (2007): México: dos ríos se desbordan dejando 400.000 damnificados. In: *ADN Mundo*, 31 October. Available from http://www.adnmundo.com/contenidos/politica/mexico-tabasco-inundaciones-rios-desbordados-lluvias-afectados-pi-311007.html, 3 June, 2008.

Andreas, P. (2000): *Border Games: Policing the U.S.–Mexico Divide*. Cornell University Press, Ithaca, NY.

Calavita, K. (1992): *Inside the State: The Bracero Program, Immigration and the INS*. Routledge, Chapman and Hall, New York, NY.

Campbell, D., and L. Berry (2003): *Land Degradation in Mexico: Its Extent and Impact*. Commissioned by Global Mechanism with support from the World Bank. Available from http://www.fao.org/Ag/AGL/swlwpnr/reports/y_lm/z_mx/mx_doc/mxtx511.doc, 15 June, 2008.

COLEF/CONAPO. (2006): *Principales Indicadores de los Levantamientos de la EMIF (1993–2003)*. Available from http://www.colef.mx/Investigacion/Documentos/Principales%20indicadores%20de%20los%20levantamientos%20de%20la%20EMIF%201993-2003.xls, 6 June, 2008.

Cornelius, W. (2001): Death at the border: The efficacy and "unintended" consequences of U.S. immigration control policy. 1993–2000. *CCIS Working Paper No. 27*, December. 2001. University of California, San Diego, CA. Available from http://www.ccis-ucsd.org/PUBLICATIONS/wrkg27.pdf, 7 June, 2008.

ECLAC. (2008): *Anuario Estadístico de América Latina y el Caribe 2007*. United Nations, Santiago de Chile. Available from http://www.eclac.cl/cgi-bin/getProd.asp?xml=/publicaciones/xml/8/32598/P32598.xml&xsl=/deype/tpl/p9f.xsl&base=/tpl/top-bottom.xsl, 5 June, 2008.

EM-DAT. (2008): *The OFDA/CRED International Disaster Database*. Université catholique de Louvain, Brussels. Available from http://www.emdat.be, 6 May, 2008.

Escobar, H., B. Sovilla, and J. López (2006): Pobreza, desastres naturales y migración en la regiones Istmo-Costa, Sierra y Soconusco de Chiapas. In: *Observatorio de la Economía Latinoamericana*, No. 70, November. Available from http://www.eumed.net/cursecon/ecolat/mx/2006/rsa.htm, 5 June, 2008.

Greenpeace. (2005): La destrucción de los ecosistemas agrava los impactos de "Stan". In: Greenpeace México Noticias, 5 October, 2005. Available from http://www.greenpeace.org/mexico/news/la-destrucci-n-de-los-ecosiste, 7 June, 2008.

Hernández, M., L. Torres, and G. Valdez. (2003): Sequía meteorológica. In: Gay, C. (ed.), *México: una visión hacia el sigla XXI. El cambio climático en México*. Mexico City. pp. 28–37. Available from http://www.atmosfera.unam.mx/editorial/libros/cambio_climatico/sequia.pdf, 10 June, 2008.Printed in 1999, online version updated in 2003.

INEGI. (2005): *Conteo de Población y Vivienda 2005*. Instituto Nacional de Estadística y de Información Geográfica, Mexico City. Available from http://www.inegi.gob.mx/inegi/default.aspx?s=est&c=10394, 20 June, 2008.

INEGI. (2000): *Censo General de Población y Vivienda 2000*. Instituto Nacional de Estadística y de Información Geográfica, Mexico City. Available from http://www.inegi.gob.mx/inegi/default.aspx?s=est&c=10252, 20 June, 2008.

de Janvry, A., E. Sadoulet, B. Davis, K. Seidel, and P. Winters (1997): Determinants of Mexico–US migration: The role of household assets and environmental factors. *CUDARE Working Paper No. 853*, University of California, Berkeley, CA. Available from http://repositories.cdlib.org/are_ucb/853/, 5 June, 2008.

Leighton-Schwartz, M., and J. Notini (1994): *Desertification and Migration: Mexico and the United States*. US Commission on Immigration Reform Research Paper, Washington, DC. Available from http://www.utexas.edu/lbj/uscir/respapers/dam-f94.pdf

Martin, P., and K. Midgley (1994): Immigration to the United States: Journey to an uncertain destination. *Population Bulletin*, September 1994. pp. 2–45.

Martinez, J. and A. Fernández (Eds.) (2004): Cambio Climático: Una Visión desde México. Secretaria de Estada de Medio Ambiente y Recursos Naturales. Instituto Nacional de Ecologica. Mexico city.

Medellín, F. (1978): *La Desertificación en México*. Instituto de Investigación de Zonas Áridas, Universidad Autónoma de San Luis Potosí, San Luis Potosí.

Nadal, A. (2000): *The Environmental & Social Impacts of Economic Liberalization on Corn Production in Mexico*, Oxfam GB and WWF International, Gland. Available from http://www.colmex.mx/programas/procientec/docsenlinea/nadal%20environmeand%20socialimpacts.pdf, 15 June, 2008.

NHI. (1997): *Breaking the Cycle: Desertification and Migration on the U.S.–Mexican Border*. Natural Heritage Institute, San Francisco, CA.

Ortíz, M., and A. Méndez. (1999): Escenarios de vulnerabilidad por ascenso del nivel del mar en la costa mexicana del Golfo de México y el Mar Caribe. *Investigaciones Geográficas* 39, 68–81. Available from http://www.igeograf.unam.mx/instituto/publicaciones/boletin/bol39/b39art4.pdf, 10 June, 2008.

Ortíz, M., and A. Méndez. (2004): Vulnerabilidad al Ascenso del Nivel del Mar y sus Implicaciones en las Costas Bajas del Golfo de México y Mar Caribe. In: Rivera, E., Villalobos, G., Azuz, I., and Rosado, F. (eds.), *El Manejo Costero en México*. EPOMEX/UACAM, Campeche. pp. 307–320. Available from http://www.uacam.mx/epomex/paginas/pdf/mancos/cap20.pdf, 10 June, 2008.

PNUD. (2007): *Informe sobre Desarrollo Humano, México 2006–2007: Migración y Desarrollo Humano*. PNUD/UNDP, Mexico City. Available from http://saul.nueve.com.mx/informes/index.html, 6 June, 2008.

Raghavan, C. (2000): Mexico: NAFTA corn liberalization fails farmers, environment. Available from http://www.twnside.org.sg/title/mexico.htm

Saldaña-Zorrilla, S. (2007): Socioeconomic vulnerability to natural disasters in Mexico: Rural poor, trade and public response. In: *Estudios y Perspectivas*, No. 92, UNO/CEPAL Disaster Evaluation Unit, United Nations, Mexico City. Available from http://www.eclac.org/publicaciones/xml/4/31414/Serie_92.pdf, 7 June, 2008.

SSN. (2007): *El Sismo de 1985 en Cifras*. Servicio sismológico nacional. UNAM, Mexico City. Available from http://www.ssn.unam.mx/website/jsp/Carteles/sismo85.jsp, 10 June, 2008.

Social Watch. (2008): *Country Report Mexico 2007*. Available from http://www.socialwatch.org/en/informesNacionales/539.html

Villafuerte, D., and M. García (2006): Crisis rural y migraciones en Chiapas. In: *Migración y Desarrollo*, 1st semester 2006, pp. 102–130. Available from http://meme.phpwebhosting.com/~migracion/modules/ve6/4.pdf, 5 June, 2008.

Wise, T., and K. Gallagher (2002): NAFTA: A cautionary tale. In: *Foreign Policy in Focus*, 24 October, 2002. Available from http://www.mindfully.org/WTO/NAFTA-Cautionary-Tale-FPIF24oct02.htm, 6 June, 2008.

World Bank. (2004): *La Pobreza Rural en México*. Available from http://siteresources.worldbank.org/INTMEXICO/Resources/La_Pobreza_Rural_en_Mexico.pdf

World Bank (2008): Gross domestic product 2008. World Development Indicators Database. World Bank, Washington D.C.

WRM. (2002): Mexico: Deforestation progresses, but not the measures to prevent it. *WRM bulletin*, No. 54, World Rainforest Movement, January 2002. Available from http://www.wrm.org.uy/bulletin/54/Mexico.html

Chapter 14
Case Study Senegal: Environmental Degradation and Forced Migration

Frauke Bleibaum

1 Introduction

Senegal, a country affected by increasing environmental degradation and the effects of climate change, experiences strong rural–urban and international migration. Field research in Central and Northern regions of Senegal described in this chapter tackles the question of the linkage between environmental factors and migration. While more than half of the migrants interviewed had to leave their home villages due to environmental degradation and problems of survival, economic and political factors do play a role in the individual's decision to migrate as well. The causes of migration identified are multiple: lack of rainfall, poor soil quality, lack of adequate agricultural policies, lack of public infrastructure in rural regions, etc. While the regions most touched by drought (Peanut Basin, stretches of the Northern Senegal River Valley) only experience out-migration, regions where irrigated agriculture can be practiced also experience immigration from other rural and urban regions. Proposals to alleviate the situation of (forced) migration include the following: adequate agricultural policies including the development of food transformation and commercialisation infrastructure, measures protecting natural resources, policies protecting land and water user rights, regional economic integration, and fair trade and agricultural policies of the countries of the North.

2 Context Information

The research reported here was conducted between January and the beginning of March 2008 in Dakar, the regional capitals Kaolack and St. Louis, and the rural communities of Ndienne Lagane, Dya, and Ross Mbethio. The research was based on expert interviews conducted with representatives and employees of state institutions, Non-Governmental Organisations (NGO), university

F. Bleibaum (✉)
University of Bielefeld, Bielefeld, Germany
e-mail: frauke.bleibaum@no-log.org

T. Afifi, J. Jäger (eds.), *Environment, Forced Migration and Social Vulnerability*,
DOI 10.1007/978-3-642-12416-7_14,

researchers, and international organisations, as well as on (semi-)structured interviews conducted with inhabitants of rural communities (RC) and the regional capitals. The aim of this research is to investigate the reasons which make inhabitants of rural regions migrate, either to the cities or to other countries. It was assumed at the outset that some of these reasons would be environmental, such as soil degradation and shortage of rainfall. The research supplies evidence for this assumption, but environmental degradation should not be regarded as the only factor making people migrate. State agricultural policies, economic conditions within the country, and changing cultural patterns also play a significant role in explaining migration of Senegalese people.

Senegal is situated in a semi-arid region where water is a scarce resource, and, as a consequence of climate change, a further decrease in rainfall in the future could occur. Due to its coastline, Senegal is also affected by rising sea levels and coastal erosion. Since large stretches of the productive land are occupied by export crops, the need for domestic food crops cannot be met, and food, especially rice, is being imported (CNCR, 2006: 10). At 2.7%, the population growth rate is higher than the growth rate of the agricultural sector, which increases the deficit in food supplies and other resources. Poverty among the Senegalese population has decreased during the last 10 years[1]; but in interviews conducted during the elaboration of the Poverty Reduction Strategy Paper (PRSP), about 65% of the households said that they considered themselves as poor (Republique du Sénégal, 2006: 7). The ethnic group of the *Wolof* makes up about 43% of the population, the *Peul* and *Toucouleur* (24%) are the second largest group, followed by others that include the *Serer* (15%), *Lebou* (10%), *Jola, Mandinka, Maures, Soninke*, and many smaller communities (Lachenmann et al., 2006).

Since the implementation of liberal adjustment policies in the early 1980s, the Senegalese government has withdrawn from its coordinating and executing activities in the agricultural sector. Before, the government had organised the production and commercialisation chain, mainly for the peanut crop, supplying farmers with fertilisers and seeds and guaranteeing the prices for their harvest. Since then, the market has been increasingly liberalised, opening new economic possibilities for private companies, but making the situation of the farmers difficult (CNCR, 2006). In 1996, certain political decision-making competences were transferred to the rural communities, such as land distribution, preservation of natural resources, and health and education issues. However, the decentralised administration (Rural and Regional Councils) generally lacks resources.

Senegal is an ancient country of immigration. During the colonial period, French, Libyan, and people from other West African countries came to the Peanut Basin and Senegalese cities as seasonal agricultural workers or tradesmen. The severe droughts that Senegal experienced in the 1970s and 1980s led to

[1] Absolute poverty touched 57% of the population in 2002, compared to 67% in 1994.

the first important internal population movements of farmers to the cities. In the last decennia, international migration to Europe, to the United States, and to other Southern and Central African countries became increasingly important. Most international migrants engage in unqualified work or in the informal sector; only some – intellectuals, students, etc. – are able to access quality jobs.

3 Environmental Degradation

The regions of the Peanut Basin and the Senegal River Valley were selected for the field study, since both regions experience strong out-migration and environmental problems, such as lack of rainfall, poor soil fertility, and the salinisation of the soil. One of the most significant climatic variations in the Sahel region has been the persistent decline in rainfall since the late 1960s. Since 2000, rainfall has not declined, but it has not reached the same level as in the 1950s either. In Senegal, the average rainfall increases from North to South; while one can have up to 1,000 mm/year in the Casamançe, the yearly rainfall in Northern regions sometimes only reaches 200 mm. The rural population often lacks the most vital resources, and, as in a vicious circle, overuses natural resources, such as wood or pasture land, which lead to an even higher poverty of those dependent on these resources.[2] Wood resources, for instance, are overexploited due to the high demand for firewood, extensive agricultural practices, and bush fires. To be able to cope with these problems, the rural communities have developed so-called Local Conventions on the Use of Natural Resources, but these often do not find sufficient recognition at the local level. Although in theory each village habitant should be able to demand land at the Rural Council (RCl), the land distribution is often organised in an undemocratic manner, thus increasing the farmer's vulnerability (Lachenmann et al., 2006: 7, 43). When farmers or stockbreeders cannot live on their resources anymore, they often have no other choice than to leave for the cities or abroad.[3]

3.1 Peanut Basin

In the regions of Fatick and Kaolack, the so-called Peanut Basin, the long history of monoculture has contributed to soil degradation and erosion, since once the peanut is harvested, the soil is no longer protected against wind and sun. Only certain crops can tolerate the high salt content of the soil and farmers often do not know what to grow anymore. One of the main problems in the Peanut Basin is the water management; rainfall is usually insufficient and the

[2] Interview Mr. Moussa Sall, State Ecological Survey Centre.

[3] Interview Mr. Seck, Environmental Development Action in the Third World (ENDA).

groundwater level is too deep to be reached easily. In recent years, not only the peanut harvest but also the harvest of subsistence crops like millet and beans has declined in the whole region (Centre de Suivie Ecologique et al., 2000). Till today, the peanut culture dominates, although the price for peanuts has dropped so that many farmers cannot live from the harvests anymore.[4] The necessary diversification of the crops needs time, since other factors like the food consumption habits and the commercialisation system play a crucial role as well.[5] Fifteen farmers, herders, and their wives in two villages in the Peanut Basin were interviewed in the field study. The farmers mentioned not only the lack of rainfall as their main problem, followed by the poor soil fertility and the salinisation of the soil, but also the lack of pasture and cultivable land. Many of the interview partners also explained that they lack the resources to buy agricultural material, seeds, and fertilisers. They stressed that the government does not support them and their activities enough. Many farmers believe that their living situation was better before the disengagement of the state in the cultivation and commercialisation of the harvest.

3.2 Senegal River Valley Region

The origins of the River Senegal are in Guinea and Mali. Downstream, the river Senegal forms the Senegal–Mauritania border before entering the Atlantic Ocean. Traditionally, livelihoods in Northern Senegal were based on pastoralism and seasonal transhumance. Annual floods of the Senegal River allowed for flood recession agriculture, which has been practiced for centuries. Flooding usually occurred between July and November in the Middle valley, an alluvial plain reaching from Dagana to Matam, and in the river's Delta (Lautze and Kirshen, 2007). The valley's agricultural production followed the river's rhythm; rain-fed cropping was practiced on the upland and flood recession farming and stockbreeding in the lowlands. Although the productivity of farming was low, the low work and capital input made this exploitation system profitable. However, the creation of boreholes in the 1950s led to an increasing sedentarisation of the population and to competition over valuable irrigated land (Cotula, 2006: 24, 38). In the 1970s, the Senegal River Basin Authority (OMVS), counting as its members the states Senegal, Mali, and Mauritania, began to organise the exploitation of the land along the river. The aims of the OMVS were the provision of secure and steadily improving livelihoods for the inhabitants of the River Valley and the safeguarding of the ecological balance of the River region (Adams, 2000:5; OMVS, 2002). Measures allowing for after-flood farming, the prevention of floods, and the just distribution of the water resources are stressed in the

[4] Interview Mr. Dia, Conseil National de Concertation er de Cooperation des Ruraux (CNCR).

[5] Interview Mrs. Willems, Vredeseilanden (VECO, Belgian NGO).

Organisations' Charter (Organisation pour la Mise en Ouvre du Fleuve Sénégal, 2002: 7). The dams constructed served as a provider of hydroenergy (Manantali Dam), as a blockage of sea water in the rivers' mouth (Diama Dam), and as a drinking water reservoir (SOFRECO, 2007: 2). The Société d'Aménagement et de Exploitation du Delta (SAED), a semi-state organisation in charge of the exploitation of the River Valley, began to establish small-scale irrigated perimeters on village level.

The advantage of the agricultural region of the River Valley, compared to the Peanut Basin, lies in the fact that irrigation allows for farming throughout the year and for a diversification of crops. Nevertheless, many factors hinder the exploitation of the valley for the benefit of the local population. During the first years after the dam construction, the artificial flooding either did not take place or water had been released several times, so that the first seeds drowned (Adams, 2000: 11). Due to short and intensive rainfall and the mismanagement of the dams on the River Senegal, there have been inundations at St. Louis and in the region[6] (Diagne, 2007). Furthermore, available pasture land was reduced significantly and the flood duration was often not sufficient to allow for the deposition of nutrient-rich sediments on the floodplains (Degeorges and Reilly, 2006: 636; Projet Biodiversité, 2007). When – under structural adjustment policies – state subsidies for farmers were cut and flood recession land was used for large-scale irrigation farming, many farmers did not have the necessary resources to continue farming on their plots. The high costs of the irrigation infrastructure make small-scale farmers particularly vulnerable to changing climatic conditions and fluctuations in harvest, since they can be deprived of the land if they do not pay the fees (Cotula, 2006: 31). According to Adams (2000: 14), the Senegal River development schemes have not brought about development but rather the destruction of production systems which provided a minimum of food security. Another problem is the commercialisation of the farmer's harvest; for instance, the locally grown rice is not sold well until today due to food consumption habits and problems of commercialisation.[7]

The farmers interviewed in one village in the River Delta confirmed that their main problems are the high costs of irrigated agriculture – the fees for the use of the irrigation system and the costs of fertilisers, seeds, and the rental of machines – and an unfair distribution of cultivable land. Some of the farmers believe that the construction of the dams along the river Senegal has not improved the farmers' situation, while others stated that they consider the constant availability of water to be less risky for farming and therefore an advantage.

---

[6] In 2003, a whole village near Podor was inundated and had to be transferred to another zone; interview Mr. Touré, SAED.

[7] Interview Mr. Touré, SAED.

4 Rural Exodus and Migration

Since the time of the disengagement of the state in the agricultural sector, most farmers, stockbreeders, and their families have been trying to leave for fertile regions or, more often, the cities. In some cases, also city habitants leave for fertile regions to engage in agriculture there[8] (Goldsmith et al., 2004). One can distinguish between two types of migration; either the whole family migrates to another region to farm there or certain family members, usually the young people, migrate to look for (seasonal) work elsewhere. Most young Senegalese – also those interviewed in this study – believe in migration to Europe as a relatively easy possibility to earn money and support their families at home. Potential migrants are being convinced by those who have left, and who, even though they have faced difficulties, come back with financial resources and pretend that everything went well.[9] Social and family relations have to be taken into account as well when explaining migrants' motives. International migrants usually do not belong to the most underprivileged groups of society. The fact that people who earn a salary in Senegal will support 30–60 other family members can function as a pushing factor for migrants, since abroad they are able to live without facing the expectations of those depending on them.[10] People who do not dispose of many resources will leave for the next town and – once they have accumulated some resources – for a bigger city or abroad. Studies on Senegalese migrants leaving for Europe by boat show that most of the passengers are young men belonging to the ethnic group of the Wolof, followed by the groups of the Serer and the Peul (SOFRECO, 2007: 6). Most Senegalese migrants to Europe can be found in France, Italy, and Spain, partly due to the networks of one of the Islamic brotherhoods in Senegal, the Mouride brotherhood. The most recent wave of migrants also includes the urban elite, since, after decades of structural adjustment and severe constraints on hiring in the public sector, well-educated young people rarely find employment in Senegal's formal sector (Riccio, 2001: 584 pp). Thus, both city and rural inhabitants opt for migration as a survival strategy. Of all 17 experts who were interviewed, 15 believe that the link between environmental degradation and migration exists, although – according to them – other factors might contribute to rural exodus and migration as well. Seven of these interview partners mentioned the lack of adequate agricultural policies at national and international levels as a factor increasing migration.

4.1 Migration Patterns in the Peanut Basin

The two rural communities in the Peanut Basin show different patterns of out-migration. Whereas Dya, a village in the region of Kaolack, witnesses

[8] Interview Mr. Maisar Fall, consultant for several farmers' organisations.

[9] Interview Mr. Dia, CNCR.

[10] Interview Mr. de Boeck, IOM.

considerable seasonal migration of young men, many inhabitants of the village Ndienne Lagane have left for Touba,[11] Dakar, or Europe on a long-term basis. Ndienne is predominantly Wolof and Mouride and since – over the past decades – Mouride networks have spread from the rural areas to the towns and to European countries like Italy or Spain (Riccio, 2001: 586), it is not surprising that many inhabitants of this village have left for Europe. Another factor explaining these migration patterns is the fact that Ndienne is, comparatively, more advantaged than other villages in the Peanut Basin. Therefore, inhabitants of Ndienne will rather have the necessary resources to migrate. According to one member of the Rural Council, the international migrants of Ndienne have left for environmental reasons. He and two other young men said that they will not leave their home village, since they could contribute to its development. All interview partners in the villages of this region – except for two – stressed that they would like to stay in their village, as long as they could live from their activities, but would leave if this was no longer possible. Similarly, of the nine migrants interviewed in the regional capital of Kaolack, only three – an artist, a craftsman, and an electrician – said that they could not imagine themselves working as farmers and would therefore not want to go back to their home villages. The others explained that they were forced to leave their home, since they could no longer make a living on agriculture and stockbreeding.

4.2 Migration Patterns in the River Valley Region

The region of St. Louis and the Northern stretch of the region of Tambacounda have the longest tradition of emigration in Senegal. Already in pre-colonial times, members of the ethnic group of the Soninké migrated from the River valley to other Sahel countries as tradesmen and, during the French colonisation, especially members of the Halpulaaren went to France as tradesmen or to study. But when the French took over the Sahel trade, many former tradesmen were forced to settle down, engaged in agriculture, or opted for (seasonal) work migration. Some 40 years ago, migration on a long-term basis to the Ivory Coast or France started to surpass seasonal migration (Lanly, 1998: 9; Ba, 1998: 95). Since the droughts of the 1970s and 1980s, the earlier tendency is partly reversed, as people from Senegal's central regions have come to work in irrigated agriculture in the River Valley. Here, farmers who have resources, often former migrants or people coming from other regions, can easily obtain fields which local workers exploit for them. But consequently, local people – especially young people and women – have difficulties in obtaining land and often see no other possibility than to migrate. Only in the middle region of the valley near Podor and in the Delta is the need to leave less persistent, since many irrigated parcels already exist there. In one village in the River Delta where I

[11] The capital of the Mouride brotherhood.

spoke with several farmers, they told me that they will stay in their village and do not want to leave for the cities. Irrigation agriculture is possible for those who can afford the inputs and the fees for the infrastructure. With their fields of rice, corn, and other vegetables, they are able to earn their living, even if with difficulty.

5 Perspectives on Rural Development and Migration

We have seen that Senegalese farmers in the mentioned regions face several problems. Among these are not only problems arising from environmental degradation but also problems of organisation and inadequate agricultural policies. Some careful conclusions can be drawn on the basis of this field work. It can be assumed that about two-thirds of the migrant interview partners were forced to leave their home villages not only due to environmental degradation and the lack of rainfall but also due to inadequate infrastructure; it remains debatable whether they should be regarded as environmentally forced migrants (Renaud et al., 2007: 29) or whether it is rather a combination of different – man-made and natural – factors which made them migrate. Since it proves difficult to separate these problems empirically, I would conclude that environmental degradation does play a crucial role, together with other related factors, in the farmer's decision to leave.

In the River Valley, where sufficient water is available, irrigated agriculture could offer an alternative to migration as soon as it proves to be possible and profitable for (more) small-scale holders. But in the Peanut Basin, farmers face decreasing soil fertility, lack of rainfall, and no possibility to irrigate their fields. This situation will lead to an even increasing rural exodus which in turn aggravates the living situation in Senegal's cities. Several proposals to alleviate the situation of forced migration can be made. Since in Senegal, an intensive competition around water and land takes place, tenure rights of small-scale holders must be guaranteed and land distribution democratised (Cotula, 2006: 72). As my interview partners said, small-scale agriculture must move into the focus of national politics again, access to land and seeds should be guaranteed, and access to water facilitated, wherever possible (Amanor, 2005; Direction de l'Environnement et des Etablissements Classés, 1999).

On the international level, several questions influencing the development of Senegal's agricultural sector and migration patterns need to be addressed as well. The International Organisation on Migration (IOM) tries to identify ways in which migrants could become agents of development by transferring knowledge or by creating employment possibilities. However, several factors hinder the possibility of migration as a contribution to national development. One notices that the migrants' remittances decrease in amount and regularity over time and that they are not invested productively within the national economy. In the case of migrants leaving for Europe, it is necessary to encourage them to come back after 1 or 2 years and to share their knowledge and competences with those who stayed behind (Mr. de Boeck of the IOM, personal communication). Nowadays, Europe usually absorbs either low-skilled workers in those sectors where they are needed or highly qualified personnel in a very specific manner. This in turn leaves the sending countries without their work force and their qualified people. According to de Boeck, the perspective on (irregular) migration has focused on Euro-centric and short-term strategies for too long; what is needed is a long-time perspective integrating inter-continental migration.

Scholars and politicians argue that migration could only be stopped by promoting social and economic development in the sending countries (African Gender Forum, 2007). However, this could only be the first step in a long-term process, since in a wealthier society, more people will be able to afford the costs of migration (Haas, 2007: 828). Only with fair trade and development policies planned on a long-term basis, will one be able to promote a global economic and political equilibrium and thus decrease forced migration. Trade and economic integration within West Africa can be another important step towards less economic vulnerability of African economies.

Besides international trade and migration policies, the agricultural and economic policies of the countries of the North also need to be addressed. One of the most important reasons pushing African people to migrate is the effects of the European agricultural policies. Europe subsidises its own farmers and stockbreeders, exports its agricultural excess products to African countries, and prevents migrant-sending countries from exporting their agricultural and industrial products. This practice destroys the national agricultural production and impedes economic growth in African countries. A fruitful option to strengthen development initiatives in African countries would therefore be the opening of the markets of the countries of the North to imports from Africa (Haas, 2007: 932).

One of the most important finding of this research, as I perceive it, is the fact that almost all of my migrant interview partners said that they would like to return to their home villages, if it were possible. Those who still live in their villages said, without exception, that they would like to stay, as long as they could earn a living from farming. One can therefore conclude that good conditions in the Senegalese agriculture and stockbreeding sector, the creation of other income-generating activities in rural areas, and fair international economic relations, including (West African) regional economic integration, could mean a control of rural exodus and migration in Senegal.

Acknowledgements I thank all my interview partners in Senegal, both those working for institutions and those having experienced and still experiencing environmental degradation and migration, for their help and support. This research was carried out within the framework of the EACH-FOR Project with the financial support of the European Commission's Sixth Framework Programme from January to April 2008 (contract number 044468). I thank EACH-FOR project colleagues for their professional support and useful critique.

References

Adams, A. (2000): *The Senegal River: Flood Management and the Future of the Valley*. International Institute for Environment and Development, Issue Paper/Drylands Program, No. 93/2000, London.

African Gender Forum (ed.). (2007): *African Women for Africa's Development – Building Partnerships with the Diaspora*. Conference held in December 2007, Dakar.

Amanor, K.S. (2005): Agricultural markets in West Africa: Frontiers, agribusiness and social differentiation. *Institute of Development Studies* 36(2): 58–63, June, 2005.

Ba, C.O. (1998): Migrations régionales et relations de genre dans la vallée du fleuve Sénégal. *Africa Development* XXIII(3 and 4), 95–119.

Centre de Suivie Ecologique; US Agency for International Development; US Geological Survey (eds.). (2000): *Monitoring Climate and Human Impacts on the Vegetation of Senegal: Drought, Agricultural Expansion, and Natural Resource management*. Eros Data Center, South Dakota/USA. April 2000.

Conseil National de Concértation et de Coopération des Ruraux (CNCR) (eds.) (2006): *Memorandum sur la situation difficile des ruraux sénégalais*. Dakar, July 2006.

Cotula, L. (ed.) (2006): *Land and Water Rights in the Sahel. Tenure Challenges for Improving Access to Water for Agriculture*. International Institute for Environment and Development, Issue paper/Drylands program, No. 139/2006, London.

Degeorges, A., and B.K. Reilly. (2006): Dams and large-scale irrigation on the Senegal River: Impacts on man and the environment. *International Journal of Environment Studies* 63(5), 633–644.

Diagne, K. (2007): Governance and natural disasters: Addressing flooding in Saint Louis, Senegal. *Environment and Urbanization* 19(2), 552–562.

Direction de l'Environnement et des Etablissements Classés. (1999): *Stratégie Nationale Initiale de Mise en ouvre de la Convention Cadre des Nations Unies sur les Changements Climatiques*. Ministère de l'Environnement et de la Protection de la Nature, Dakar.

Goldsmith, P., K. Gunjal, and B. Ndarishikanye. (2004): Rural–urban migration and agricultural productivity: The case of Senegal. *Agricultural Economics* 31, 33–45.

de Haas, H. (2007): Turning the tide? Why development will not stop migration. *Development and Change* 38(5), 819–841.

Lachenmann, G., F. Bleibaum, J. Ehlert, L. ElOumrany, D. Krenz-Dewe, G. Lachenmann, F. Odoemenam, Y. Ogawa-Müller, S. Vennemann, and B. Zohy (2006): Décentralisation, société civile, développement au Sénégal. *Working Paper No. 357*, Universität Bielefeld, Bielefeld.

Lanly, G. (1998): *Les immigrés de la vallée du fleuve Sénégal en France :des nouveaux acteurs dans le développement de leur région d'origine*, Division du Développement Rural de l'Organisation des Nations Unies pour l'Alimentation et l'Agriculture (FAO), Dakar.

Lautze, J., and P. Kirshen. (2007): Dams, health and livelihoods. Lessons from the Senegal, suggestions for Africa. *International Journal of River Basin Management* 5(3), 167–168.

Organisation pour la Mise en Ouvre du Fleuve Sénégal (OMVS). (2002): *Charta des eaux du fleuve Sénégal*. Dakar.

Projet Biodiversité Mauritanie – Sénégal. (2007): *La biodiversité dans la vallée du fleuve Sénégal*.

Renaud, F., J. Bogardi, O. Dun, and K. Warner (2007): *Control, Adapt of Flee – How to Face Environmental Migration?* UNU-EHS, Interdisciplinary Security Connections no. 05/2007.

Republique du Sénégal. (2006): *Document de Stratégie de Croissance et de Réduction de la Pauvreté 2006–2010*. Dakar.

Riccio, B. (2001): From ethnic group to transnational community? Senegalese migrants' ambivalent experiences and multiple trajectories. *Journal of Ethnic and Migration Studies* 27(4), 583–599.

SOFRECO. (2007): *Les nouvelles Tendances de l'émigration au Sénégal*. Cedex.

Chapter 15
The Environmental Root Causes Triggering Economic Migration: The Case of Egypt

Tamer Afifi

1 Introduction

Egypt covers a total area of about 1 million km^2, most of which experiences arid and hyper-arid climatic conditions. The six main agro-ecological zones in Egypt are the following (Arab Republic of Egypt Ministry of Agriculture and Land Reclamation, 2002):

1. The Nile Valley, including the fertile alluvial land of Middle and Upper Egypt, where the main source of irrigation water is the Nile River. Agriculture production of Egypt is mainly concentrated in this zone in addition to the Delta.
2. The Nile Delta region, where the main source of irrigation water is the Nile River as well. Together with the Nile Valley, the agriculture production in this zone covers about 6.6 million acres. Most of the soil in both areas is recent Nile alluvium.
3. The reclaimed desert areas on the fringes of the Nile Valley, where the only source of irrigation is the groundwater.
4. The North Coastal zone, including the coastal area starting from the North-Western coast moving eastwards to North coastal area of the Sinai Peninsula, where there are no reliable figures available on groundwater quantity and usage.
5. The Inland Sinai and the Eastern Desert, where the main source for irrigation is the groundwater.
6. The Western Desert, including oases and southern remote areas, where the groundwater is mainly extracted from the Nubian Sandstone and carbonate aquifers.

T. Afifi (✉)
Environmental Migration, Social Vulnerability and Adaptation Section, United Nations University Institute for Environment and Human Security (UNU-EHS), Bonn, Germany
e-mail: afifi@ehs.unu.edu

T. Afifi, J. Jäger (eds.), *Environment, Forced Migration and Social Vulnerability*,
DOI 10.1007/978-3-642-12416-7_15,

Egypt is particularly vulnerable to climate change because of its dependence on the Nile River as the primary water source, its large traditional agricultural base, and its long coastline, already undergoing intensifying development and erosion.

Ninety-seven percent of Egypt is desert; only 5% of the land area in Egypt is actually occupied and less than 4% of the land is suitable for agriculture. Because such a small percentage of land is habitable, population densities in these areas (that is, the coastal zones and along the Nile River) are high, and in some areas along the Nile River even greater than 1,000 people per km^2. The coastal zone is home to more than 40% of the population; this zone is particularly vulnerable to climate change, where the main challenges are sea-level rise, shoreline erosion, stresses on fisheries, and saltwater intrusion in groundwater (Arab Republic of Egypt Ministry of Agriculture and Land Reclamation, 2005).

The powerful Nile River nourishes a slim column of Egyptian territory, but the remaining part of the country faces extreme water scarcity. The majority of the huge population (80,335,036 in 2007) (CAPMAS, 2008) in Egypt uses the Nile as the main source of water. This huge population puts pressure on the limited natural resources and contributes to the poor human conditions of different demographic groups. For example, Egypt is ranked 116th out of 179 countries in the Human Development Index (UNDP, 2008).

This chapter attempts to identify the most significant environmental problems in Egypt and link them to migration within the country and to other countries. The chapter focuses on desertification in its broad sense (including soil degradation, soil salinity, soil erosion, sand dunes) as well as water shortage and their impact on migration. However, the chapter also sheds some light on other environmental problems in Egypt, such as air pollution and sea-level rise, without going into much detail, due to time and budget limitations during the field work, and due to the fact that a phenomenon like sea-level rise is not yet a top priority on the agenda of many policymakers, and most of the inhabitants are not yet aware of the problem as shown in the next section. The chapter starts with a brief overview of the environmental problems in Egypt, moving on to the migration processes in the country, then demonstrating the methods used in the field research. After the findings are shown, conclusions are drawn from the field work analysis.

2 An Overview of the Environmental Problems in Egypt

Air pollution and sea-level rise are two important environmental problems the Egyptians have to face. The first problem has been living with the Egyptians for decades, especially in Cairo, the capital. The huge increase in population led to a consequential surge in the number of vehicles, with an associated significant increase in street traffic, exceeding 3 million vehicles in 2004. On the other hand, sea-level rise is a future challenge that threatens millions of inhabitants and has not yet had its direct influence on the people. This ‘postponed’ environmental problem – or rather time bomb – is not yet considered on a broad scale among the Egyptian government officials, although a study on the vulnerability of the coastal zone in Alexandria, the second largest city in Egypt, revealed that a 30-cm rise in sea level is expected to occur by 2025, inundating approximately 200 km^2. As a result, over half a million inhabitants may be displaced and approximately 70,000 jobs could be lost (El-Raey et al., 1999). An estimated 50-cm rise in sea level could affect one-third of Rosetta's[1] employment with significant economic losses estimated in billions of dollars (El-Raey, 1997).

2.1 *Sand Dunes*

The encroachment of sand and sand dunes occurs mainly in the Western desert towards fertile sedimentary land; an estimated area of 16% of the old valley land has been influenced. Satellite images reveal the size of this impact. The area influenced by this phenomenon is estimated to be 1.6 million Feddans.[2] The area influenced by the active encroachment of sand and sand dunes is estimated to be 1.6 million hectares. Land productivity has decreased by about 25% compared to its productivity in the 1960s (ACSAD, 2000).

2.2 *Waterlogging and Soil Salinity*

Due to the wide use of flood irrigation for water usage, excessive flood irrigation from the Nile is leading to soil waterlogging and poor drainage of excessive water that exceeds the growing plants' needs. Soil salinity reaches a level that damages plant production due to the deterioration of some of the chemical and biological soil elements. Of the Northern cultivated land and both Middle and Southern Delta regions, 60% and 20%, respectively, are considered to be salt-affected soils (SRA, 2000). In order to compensate for the low productivity caused by soil salinity, the usage of organic fertilizers, the adoption of inappropriate agricultural management techniques, and excess irrigation increased, which enhanced the problem and created a vicious circle. In concrete numbers, around 30% of the irrigated farmlands in Egypt suffer from salinity.

2.3 *Urban Encroachments*

Urban encroachment occurred due to the expansion of cities and villages and the establishment of industrial facilities and infrastructure, in addition to soil surface scrapping for manufacturing red bricks. Soil scrapping has been nearly overcome as a result of the legislation issued in 1983 and amended in 1985. Other encroachments started during the 1950s causing the loss of 15,000

---

[1] Rosetta is one of Egypt's historical cities.

[2] Feddan is an Egyptian/Arabic scale unit for measuring land size. 1 Feddan = 24 Kirat = 300 Kassabas = 4201 m^2 (0.42 ha).

Feddans annually (SRA, 2000). A military order was issued to stop and eliminate such encroachments in 1996, significantly limiting them, but probably after it was too late.

2.4 Land Degradation

Many nutrient elements of the Nile Valley and Delta soil were depleted by the extensive and frequent cropping, unsustainable irrigation water management, and improper agricultural practices. The construction of the Aswan Dam in the South of Egypt decreased the annual additions of the fertile sediments to the soils that consequently lost much of their content of organic matter, total nitrogen, and other nutritive elements.

With the beginning of the 1970s, soil pollution started due to the excessive use of chemical (nitrogenous) fertilizers after the construction of the Aswan Dam (SRA, 2000). Excessive use of chemical fertilizers persisted due to agriculture production intensification and attempts to reach the highest production possible per unit area. Soil and water pollution has also increased due to the inappropriate use of different pesticides. In addition, wastewater and industrial drainage leakage into watercourses has exacerbated soil and water pollution, especially since there was very poor implementation of pollution control regulations.

2.5 Water Shortage

Since the dawn of history, a continuing struggle between the Nile and Egypt's inhabitants has taken place. As the Egyptians realized the importance of the Nile for their existence long ago, they made a concerted effort to regulate the use of its water and control its floods.

Dams in Egypt have served people for at least 5,000 years. History does not record exactly when irrigation systems and dams were first constructed. The very first reported attempt was about 3400 BC and the Aswan High Dam was constructed in the late 1960s (Ouda, 1999).

In spite of all the Agreements and plans related to the Nile river, and as much as the former has been a generous water resource for Egypt, the Egyptians are suffering from water shortage due to the continuous increase in population. Taking the expression 'water shortage' more broadly, it would encompass the access to clean water that is suitable for drinking and irrigation; unfortunately, Egypt has been notorious for water pollution, since the Nile and its canals have been subject to industry, agricultural, and individual wastes. Poor water management due to inefficiency of the traditional gravity irrigation system, inadequate maintenance of irrigation and drainage networks, and over-abstraction of groundwater, especially in the newly reclaimed desert areas, are all factors that magnified the problem. Another natural factor that diminishes the available fresh water is the water salinity, a phenomenon that largely exists in the newly reclaimed desert lands that rely on groundwater.

3 An Overview of the Migration Processes in Egypt

Unfortunately, there is a big lack of data about internal migration in Egypt. The reason for that could be the fact that many Egyptians move from one region/city to the other within 1 day, due to work obligations, and therefore, it is hard to calculate the exact number of people who moved and really settled in the different cities. Moreover, in contrast to the internal movements it is easier to monitor the number of people who crossed the country borders, since all the movements are documented.

The Egyptian constitution considers emigration as a basic right for the Egyptian citizens. This was confirmed by Article 52 of the permanent constitution released in the year 1971 and stating that: '... The Egyptian citizens are entitled to emigrate permanently or temporarily...' (Arab Republic of Egypt Ministry of Manpower and Migration).

According to the official estimates of the Central Agency for Public Mobilization and Statistics (CAPMAS), the total number of Egyptian temporary migrant labourers (in Arab countries) is about 1.9 million (CAPMAS, 2001).[3] The top five destination countries are as follows: Saudi Arabia (48.3%), Libya (17.4%), Jordan (11.9%), Kuwait (10%), and UAE (United Arab Emirates, 5%) (CAPMAS, 2001).[4]

The total number of permanent Egyptian migrants in non-Arab countries is slightly more than 0.8 million (824,000). About 80% of them are concentrated in five countries: USA (318,000 or 38.6%), Canada (110,000 or 13.3%), Italy (90,000 or 10.9%), Australia (70,000 or 8.5%), and Greece (60,000 or 7.3%). The other 20% are mainly in Western Europe countries, such as the Netherlands, France, England, Germany, Switzerland, Austria, and Spain (CAPMAS, 2000).[5]

---

[3] Although the CAPMAS census for the year 2008 is available, there are no updated figures on the migrants and immigrants after the years 2000 and 2001.

[4] The statistics given by CAPMAS are just estimates which are drawn from the reports of Egyptian embassies abroad, records of cross-border flows from the Ministry of Interior, emigration permits from the Ministry of Manpower, and some other sources. The receiving countries make different estimates than CAPMAS. For example, the Italian government estimates that there are around 35,000 Egyptians in Italy whereas CAPMAS gives a figure of 90,000.

[5] Logically, there are no concrete or reliable figures about illegal migration. Nevertheless, there are tens – if not hundreds – of Egyptians who were caught in the past years after illegally attempting to cross the Mediterranean to reach Italy, Malta, and Spain. The source of information is usually the newspapers, but there are no real statistics that cover the phenomenon. When arrested, most of them mention poverty and unemployment as main reasons. None of them – to the knowledge of the author – mentions political or environmental reasons.

4 Methods

The field trip to Egypt was based on using two methods, namely, running expert interviews (government officials, representatives of Non-Governmental Organizations [NGOs], and university professors) and distributing questionnaires among people who left their homes for other cities/regions.[6] The research was carried out within Egypt, and therefore, it was not possible to make interviews with people who left the country entirely. However, from the questionnaires filled out, one can listen to stories that these interviewees told about their relatives/friends who crossed the borders. Since at the time of the field trip in Egypt a non-migrant questionnaire had not been designed, the author tried to adapt the migrant questionnaires to the cases of non-migrants – who resisted the environmental problems and did not leave their regions/villages – using the expression 'What would, if...' in order to get as much useful and relevant information as possible.

The migrants who were interviewed were mainly in the centre of the Nile Delta, the Nile Valley (South and North), Eastern and Western Nile Delta, newly reclaimed desert lands, and slums of Old Cairo. The questionnaire was semi-structured and covered other non-environmental factors that could have led to the migration of interviewees, such as poverty and social problems. The selection of the interviewees was random, due to the limited awareness of the topic 'environmental migration' and the difficulties of accessing the target groups in an organized way.

The target groups were people who left the Nile Valley and Delta for slums of Old Cairo, people who left the Oasis and moved to Cairo, people who moved from one area to the other within the Nile Valley and Delta, people who moved from fertile lands to the newly reclaimed desert lands, and people who stayed in the Nile Valley and Delta. Due to the relatively limited number of interviewees, the author did not only rely on the information about their personal experiences but also on their stories about their parents, relatives, and friends.

Not all the questioned people had decided to migrate; some of them had to move with their parents years ago or were not even born when the move occurred, particularly in the case of the Aswan Dam construction, where the parents were officially displaced in the 1960s and 1970s or in the cases of soil scrapping and urbanization in the 1980s.

[6] Since Egypt was one of the first case studies run in the frame of the EACH-FOR project, only migrant questionnaires were designed and used. Nevertheless, the author of this case study recommended designing additional questionnaires to be filled out by non-migrants who resisted the environmental degradation and stayed in their villages/regions, since he found out during his trip in Egypt that even this category of people could be relevant to the research. The non-migrant questionnaire was then designed and used in other case studies that took place afterwards.

5 Fieldwork Findings

5.1 Expert Interviews

The total number of interviewed experts was 22 (Afifi, 2009). A number of academics in Egypt (Faculty of Economics and Political Science, Cairo University) believe that there is no real impact of environmental degradation on migration, since usually migration is induced by poverty and unemployment, especially in the cases where people leave rural areas and are attracted to urban areas where they seek more and various job opportunities and better livelihoods. At least they believe that poverty and unemployment are important factors that intervene in the migration decision, even if environment plays some role. In addition, they argue that environment-induced migration is difficult to monitor, since the most significant environmental problem in Egypt, which is desertification, is a long-term problem, and it would take generations in order for such migration to be sensed.

Most of the rest of interviewees did not deny the link and provided the author with the knowledge about 'environmental migration' as will be shown in the following.[7]

Many people migrated to two different regions in the extreme Western and Eastern Delta in the frame of the Mobarak National Project. The reasons why they migrated were mainly unemployment, poverty, and overpopulation in the regions of origin, which were mainly Cairo, Beheira, Kafr El-Sheikh, and Qalioubia. However, since these two regions of destination lack access to potable water, proper infrastructure, public facilities, schools, health care, and well-functioning sewage systems, many of them left and others are expected to leave either to other regions or to return back to their original regions. For example, the Eastern Delta Region obtains its water from the Salam Canal, which is extremely polluted.

The fisheries in the Northern Lakes (Manzala, Maryout, Edkou, and Borollos) are getting polluted due to the inflow of agricultural wastes as well as sewage water. The quality and quantity of the fish are deteriorating; only the fish that can survive water pollution remain. Since these fish are cheap, the revenues of the fishermen decrease and they have to leave for better livelihoods, looking for better quality fish in the Nasser Lake (Southern Egypt). Some leave for Cairo and change their activities. Others seek new livelihoods along the regional waters in Eritrea, Somalia, Ethiopia, and Sudan and are subject to arrest by the border police. Some of these fishers try to escape to Europe (mainly Malta and Italy).

Due to the pressure on the natural resources caused by population growth, the people are not only suffering from water shortage, but also the soil quality of

---

[7] As far as the research budget allowed, the author visited most of the regions, based on these interviews, in order to meet environmental migrants on site and fill out questionnaires.

the Egyptian fertile lands has deteriorated, especially after the Aswan Dam was constructed; the permanent irrigation has replaced the seasonal irrigation system, which further harmed the soil through the overuse of chemical fertilizers and led to more salinity as well.

In line with the general strategy in Egypt, covering the drinking water needs takes the first priority, followed by industrial and agricultural needs as the second and third priorities, respectively. Therefore, the agricultural sector suffers most from water shortage, leading in turn to migration flows to Cairo, where farmers change their activities, seeking higher standards of living.

Environment per se does not lead to migration to the Cairo slums; the economic factors that are influenced by environment degradation influence in turn the migration decision. People leave rural areas due to the declining yields caused by soil degradation and water shortage and move to the Cairo slums which are massively suffering from 'other sorts' of environmental problems, such as limited access to potable water, sunshine, and fresh air, factors leading them to leave these areas again, if they can afford it.

The sand dunes occurring in the Western Desert lead to the complete 'disappearance' of some villages, such as the old Ganah and the Moschée village within the Kharga Oasis. This results in migration flows of the original inhabitants, where there are no concrete statistics that show the number of displaced people or their destination regions. The same phenomenon occurred in the Eastern Desert, where a tribe called 'Ababda' entirely left the Eastern zone, due to sand dunes and droughts.

5.2 *Migrants Questionnaires*

Besides the expert interviews, the field trip to Egypt targeted migrants who left their home for other regions within Egypt,[8] in an attempt to find out whether environmental problems influenced their migration decision; 31 questionnaires[9] were filled out, where 26 left their home regions while 5 stayed in their home lands. At this point, it is of great importance to highlight the drawback associated with the very limited number of questionnaires filled out in the field, especially that they were geographically scattered in the country.

Answering the question about whether at any point of time environmental problems affected their decision to move, most of the interviewed migrants had a positive answer, and many of the latter took the decision on their own. The rest just followed their parents in their childhood.

Almost half of the migrants expect environmental problems in the future to make them and their families migrate to different places, and most of the latter

[8] The field trip did not cover migrants who left the country due to financial constraints.

[9] Due to time and financial constraints, the number of questionnaires was not big enough to run an econometric model.

are indeed planning to leave due to environmental problems. Interviewees who are not willing to leave as well as people who did not leave in the first place are either attached to their regions or adapted to the environmental problems they are facing.

Most of the people who left the relatively fertile lands in the Nile Valley and Delta and moved to Cairo were mainly induced by unemployment and poverty problems. However, in most cases land degradation and water shortage were actually the real causes of these problems, even if this was not mentioned explicitly by the migrants. The interviewees who mentioned that they left their Oasis in the Western Desert mainly to seek better livelihoods and standards of living in Cairo referred implicitly to the sand dunes that hindered them from planting and shepherding properly in their original home areas.

People who moved from one area to the other within the Nile Valley and the centre of the Delta had different reasons; some had moved with their parents who worked as peasants or farmers in the early 1980s when the owners of the fertile lands decided to use the latter in constructing new buildings or for soil surface scrapping for manufacturing red bricks. In these cases, their parents were displaced from the lands, since they neither owned them, nor were there signed contracts that would preserve their right to stay or – at least – to be compensated. Since the soil scrapping mentioned in a previous section mainly took place in the 1970s and 1980s, the author interviewed the second generation of the people who suffered from this phenomenon.

Others moved within the Nile Valley and Delta, since they were unemployed, but this group would not dare to entirely shift to the completely different 'life style' of Cairo. Therefore, they moved to similar areas, but where they could find better livelihoods. Among this group, some left their original towns in the Nile Valley and the centre of the Delta for Eastern and Western Delta, where the Egyptian government has initiated two major projects close to the Bahr El-Bakar Canal and the Noubaria Canal, respectively. However, since they are suffering from environmental problems, many of them are willing to move back or to new places.

Another interviewed group were the people who moved to the newly reclaimed desert lands, since they were unemployed in their home towns and were promised work as peasants in these lands. However, most of them were suffering from soil and water salinity problems. Some of them even already moved to different areas within the newly reclaimed desert lands mainly because the owners of the lands decided to sell them when they did not have enough financial means to dig for new groundwater.

A group of interviewees in the Nile Valley and Delta were suffering from water shortage and land degradation. However, since they own the land and feel emotionally attached to it, they would not leave it, even if the pull factors in the city or elsewhere were very strong. As long as they can afford their living, even if it worsened in the past years, they would stay. Specifically, for this group of people and the same category in other countries within the Environmental Change and Forced Migration Scenarios (EACH-FOR) project, the author recommended designing non-migrant questionnaires. In this case study, the migrant questionnaire was adapted so that as much information as possible could be gathered about these people.

None of the interviewees completely left the country. Nonetheless, when some of them mentioned their relatives/friends who crossed to borders, mainly to other Arab countries, they mentioned poverty and unemployment as the main reasons, even though these two factors may be caused indirectly by water shortage and land degradation.

6 Analysis of Findings

From the above sections it is clear that most of the migrants did not mention environmental problems explicitly as a cause for their migration. There are many other factors that intervene, such as poverty and unemployment, which are more striking for the interviewees, even if the poverty and unemployment are totally or partially caused by environmental degradation.

There are considerable pull factors that support the migration decision of people who are affected by environmental degradation. For example, the style of life in Cairo and the higher living standards and income in the Gulf countries are strong reasons why people leave their towns/country and travel.

People would leave their home and move to another place only if there are absolutely no more livelihood possibilities for them, such as the example of soil scrapping or urbanization, where it is not their decision but the land's owners. In this case, they have to leave for other places/regions in order to make their new living. Another case is the official displacement that occurred during and after the construction of the Aswan Dam and where the people living in Nubia had to leave for Kawm Umbu after their land was covered by the water of the Nasser Lake.

The migrants who leave their villages/regions/the whole country can only do that if they have the financial means, which is in many cases not possible, since the environmental degradation had a negative impact on their income, creating hereby a vicious circle. Usually, the farmers and cattle herders belong to poor families, and therefore, borrowing money for travelling is not a common option.

The ownership of the land is a very important fact that influences the migration decision. As long as the farmers are hired on the land, they are very mobile and flexible in response to environmental changes. Hence, there is an inverse relationship between the ownership of the land and the willingness to migrate. Nevertheless, owners of the land would not leave unless there is no other way or they are officially displaced by the government.

It is quite obvious that the state does not yet consider environmental migration to be a serious problem, or at least, the issue is not one of the important priorities. This can be concluded from the interviews with the different experts. The state is more interested in internal and international migration in general without really directly relating it to environmental degradation. For example, the projects that have been established in the Western and Eastern Delta as well as the desert reclamation are aiming to improve the general conditions of the farmers, no matter whether the problems they have been suffering from in the regions of origin are purely environmental or economic. The issue is to what extent these projects are successful; it is clear that there is still a lot to be done concerning the infrastructure in Western and Eastern Delta and the salinity in the newly reclaimed desert lands. An option could be digging canals that connect these lands with the Nile River.

It seems from the field research that the distance does not really play an important role when moving/migrating from one place to the other, as long as, first, the movement takes place within the country, second, people have the financial and social means to move, and third, the new places offer them considerably better living conditions. Evidence for the latter is that a number of the interviewees are seriously considering going back to their home towns, since the conditions of their new destinations did not meet their expectations. In this case, the distance between the new and old places has no considerable impact on the decision to leave/return back.

It is a hard task to foresee the future concerning environmental migration in Egypt, since there are many factors playing a role; as mentioned above, sea-level rise could have an enormous effect on migration. However, in the context of the field research and interviewing potentially affected people, it was not possible to tackle this problem, since people are not aware of it, and it is not clear how people would react/cope in this case. Concerning the other 'tangible' factors, such as land degradation and water shortage, it can be expected that people will keep moving from one region to the other, as long as they can afford it. If the conditions in the original homes would improve, most of them would return back, unless the reason why they migrated was rather the pull factors in the regions of destination, such as higher living standards and style of life. As for the phenomena related to forced migration or displacement, such as soil scrapping and the construction of the Aswan Dam, these events already took place in the past, and therefore, they are not relevant for future scenarios, unless new development projects are initiated, and in this case, lessons can be learned from the past. However, there is no evidence that people would not be displaced again in such cases, since there are other political and economic factors intervening when designing development projects. Talking specifically about soil scrapping, the law released in the 1980s explicitly forbids such practices, and therefore, the phenomenon has not really occurred afterwards.

7 Conclusions and Future Research

Environmental problems, such as water shortage and land degradation, are certainly important challenges facing the Egyptians, especially given the rapid population growth. From the results of the field work in Egypt it can be concluded that people in Egypt would be moving/migrating from one place to the other within the country in the case of water shortage and/or land degradation only under certain conditions; they would only move, if they are not land owners, if they can socially and financially afford to leave their place of origin, and/or if they are forcibly displaced by the government or the owners of the land. According to interviews with the friends and relatives of people who left the entire country, water shortage and land degradation as push factors would need to be supported by additional push factors in Egypt and/or pull factors in the countries of destination, in order for people to take the migration decision. Additional push factors in Egypt include poverty and unemployment, and pull factors in the receiving countries include a large financial return and/or higher living standards. In general, as long as the people are not facing a sudden natural disaster, such as earthquakes or floods, they would not be willing to leave their home.

An important question would be, What implications would the environmental problems in Egypt have on the migration to Europe? As referred to in Section 3, thousands of Egyptians leave their country each year with the intention of permanently resettling in various European countries. However, this estimate is general and does not refer particularly to the environmental migrants. Moreover, the data are based on the most recent CAPMAS census covering this issue which was released in the year 2001; since this time, the visa requirements for non-European citizens to enter Europe – and non-Americans to enter the USA – have become stricter. There is definitely a large – but unknown – number of illegal 'migrants' to Europe, but the extent to which environmental problems contributed to the migration decision in such cases remains vague, simply because the access to this category of 'migrants' is hardly possible. In the cases of Australia and New Zealand which are in geographical terms – as compared to Europe – far from Egypt, the economic and political global changes also had their impacts on migration from other countries. However, it is still easier for Egyptian citizens – regardless of the reason behind their decision – to enter and to migrate to these two countries; since the demand for foreign labour is higher than in the case of Europe, the visa requirements are still not as strict.

There are many gaps that need to be filled in the future for more comprehensive studies; the time allocated for field research should without doubt be longer, in order for the researcher to be able to run more in-depth analysis and meet different target groups. Moreover, the number of questionnaires filled out is not sufficient for running econometric analysis. Adding questionnaires of different case studies is not an option, since the different countries with their environmental problems and migration patterns as well as other factors are not homogeneous. It would also be useful to run the research over a longer period of time in order for the study to be comparative by questioning the same target groups twice with a time lag, for example, 3 years, especially given that the environmental impacts in Egypt are usually a matter of time rather than sudden natural disasters.

Acknowledgements The research was conducted as part of the Environmental Change and Forced Migration Scenarios (EACH-FOR) research project, co-financed by the European Commission – Sixth Framework Programme (contract no. 044468).

References

Afifi, T. (2009): Niger Case Study Report for the environmental change and forced migration scenarios project. Available from http://www.each-for.eu/

Arab Center for the Studies of Arid Zones and Dry Lands (ACSAD). (2000): Regional report on desertification in the Arab World. Damascus.

Arab Republic of Egypt Ministry of Agriculture and Land Reclamation. (2002), Egyptian national action program to combat desertification, Desert Research Centre (DRC), Cairo.

Arab Republic of Egypt Ministry of Agriculture and Land Reclamation. (2005): Egyptian national action program to combat desertification, Desert Research Centre (DRC), Cairo.

Arab Republic of Egypt Ministry of Manpower and Migration. (2007): Available from http://www.emigration.gov.eg/

Central Agency for Population Mobilization and Statistics (CAPMAS). (2000, 2001). Cairo.

Central Agency for Population Mobilization and Statistics (CAPMAS). (2008), Population clock, Cairo.

El-Raey, M. (1997): Vulnerability assessment of the coastal zone of the Nile Delta of Egypt to the impacts of sea level rise. *Ocean and Coastal Management* 37, 29–40.

El-Raey, M., Kh. Dewidar, and M. El Hattab. (1999): Adaptation to the impacts of sea level rise in Egypt. *Climate Research* 12, 117–128.

Ouda, A.M. (1999): *Egyptian Policy and the Nile Water in the 20th Century*. Centre for Political and Strategic Studies, Cairo.

Science Research Academy (SRA). (2000): Land, water and environment institute: The impact of reduced Nile alluvial 1971–1975.

United Nations Development Program (UNDP). (2008): Human development report (HDR). Available from http://hdr.undp.org/en/statistics/

Chapter 16
A Country Made for Disasters: Environmental Vulnerability and Forced Migration in Bangladesh

Alice Poncelet, François Gemenne, Marco Martiniello and Hassan Bousetta

1 Introduction

There is a broad consensus amongst the scientific community that South Asia is amongst the regions most affected by climate change. According to the Intergovernmental Panel on Climate Change (IPCC) Fourth Assessment Report (2007) the main climate change impacts in the region are as follows: increased frequency of droughts and floods negatively affecting local production; sea-level rise exposing coasts to increasing risks, including coastal erosion and growing human-induced pressures on coastal areas; and glacier melt in the Himalayas with more flooding and rock avalanches. Crop yields could decrease up to 30% in Central and South Asia by the mid-twenty-first century. Within South Asia, Bangladesh is the most vulnerable country because of its regional connectivity through geo-physical and hydrological features and its livelihood reliance on trade (ELIAMEP, 2008).

Most of Bangladesh sits astride the deltas of a series of large rivers flowing from the Himalayas and about a third of the country is flooded every year during the monsoon. Many people live and farm on flood-prone land. Floods kill hundreds and make thousands homeless every year.

The country is criss-crossed by 7 major rivers and over 200 minor ones. These rivers and tributaries define the geography of the country and the life style of its people. The rivers are the main sources of irrigation, the principal arteries for commercial transportation, and homes for fisheries and aquatic resources. This makes the communities living along the riverbanks predominantly dependent on the flood plains for their livelihoods and forces them to be more exposed to natural hazards like floods, cyclones, and droughts. The vulnerability is enhanced by problems like upstream river erosion and climate change (CBDM, 2008). Since it is a delta country, most parts of Bangladesh are less than 12 m above the sea level, and it is projected that about 50% of the land

A. Poncelet (✉)
Centre for Ethnic and Migration Studies (CEDEM), University of Liege, Liege, Belgium
e-mail: poncelet.alice@gmail.com

T. Afifi, J. Jäger (eds.), *Environment, Forced Migration and Social Vulnerability*,
DOI 10.1007/978-3-642-12416-7_16,

would be flooded, if the sea level were to rise by 1 m (Ali, 1996). Consequently, floods occupy a unique position in the economy and culture of Bangladesh. Local people know how to live with regular floods, but their livelihoods are now at threat because of the ever-growing impacts of climate change.

This chapter aims to assess the importance of environmental drivers in the current rural–urban migration flows. From this assessment, the migration behaviour of the population, when confronted with environmental change, is described. Particular attention is given to the thresholds that might induce mass migration as well as migration patterns.

The first section provides a general overview of the patterns of environmental changes in Bangladesh. The profile of those who decided to migrate and resettle as well as those who were willing – but not able – to migrate is then analysed. Building upon these contrasted descriptions, a tentative typology of environmental migrants is discussed. This section also explores the different impacts of these migratory movements on the social, political, and economic structures of the country. The third and final section is devoted to an analysis of migration and resettlement policies implemented by the Bangladeshi Government to manage and cope with these population movements.

The chapter is based on fieldwork conducted in Spring 2008 within the framework of the Environmental Change and Forced Migration Scenarios (EACH-FOR) project. It is based not only on expert interviews conducted mainly in the capital, Dhaka, but also in the Gaibandha, Satkhira, and Khulna districts. Research was also undertaken directly with migrants and affected people from the South-West and from the North-West of the country as well as others living in the slums of Dhaka. This part of the research was based on interviews and standardised questionnaires.

2 An Overview of Environmental Changes in Bangladesh

The total area of Bangladesh is merely 144,000 km^2 with a population of about 150 million, making it one of the most densely populated countries in the World (1,050 inhabitants/km^2). As mentioned above, Bangladesh is also one of the most environmentally vulnerable nations with regard to climate change. The environmental problems of Bangladesh, such as poor soil, water, and air quality, in addition to natural disasters like floods, droughts, and tropical storms make it difficult for the people to survive in the places where they live.

The country is situated in a fertile alluvial plain where rice, tea, jute, and other crops are grown. The economy is primarily based on land use, with the majority of the country's population in agricultural employment. The average land holding is about 0.11 hectares (Mahtab and Karim, 1992:50).

It is often said that Bangladesh is a country made for natural disasters. The geography and climate have made Bangladesh prone to periodic floods, river erosion, and cyclones. Too much rainfall during the monsoon season (May–October) causes large-scale floods leading to farmland crop losses and loss of properties and human lives. In addition to flood disasters, riverbank erosion caused by the continuous shifting of channels within the major rivers Jamuna, Ganges, and Meghna is an endemic problem that displaces, according to Disaster Management Bureau (DMB) sources, an estimated 500,000 people annually and creates much distress in the country. Riverbank erosion is considered to be one of the principal contributors to the process of impoverishment and marginalisation of rural families due to the loss of productive agricultural lands.

2.1 Riverbank Erosion

About 80% of the Bangladeshi people live in rural areas and less than 40% own cultivable land. A major cause of landlessness is the riverbank erosion. The impact of riverbank erosion is exacerbated by floods and other natural calamities. The shifting of the river channel causes land erosion (Hossain, 1991: 307). People are forced to leave the affected areas, that is, forced migration occurs due to the catastrophic changes in the environment. About 94 upazilas[1] of 50 districts face riverbank erosion each year (ACD, n.d.). Furthermore, about 135,632 families became homeless due to riverbank erosion on 12 rivers, including the three major rivers in the last 5 years (ibid.). The main three rivers of the country, Padma (also called Ganges), Meghna, and Jamuna (called Brahmaputra in India), as well as another 16 rivers, are the main locations of riverbank erosion.

At the time of the monsoon, one-third of the territory on average is flooded in normal times. Residents have learned to adapt to these climatic issues, but global warming has broken this mature pattern. On the one hand, the increase in rainfall during the monsoon, coupled with the melting of Himalayan glaciers, is increasing the quantity of floodwater. On the other hand, sea-level rise makes the streamflow increasingly difficult. The annual floods are expected to gain gradually in volume and duration.

In Bangladesh, a significant number of inhabitants are victims of riverbank erosion and floods every year. The impact of riverbank erosion and of unusual floods on the local economy is great as it erodes agricultural land, infrastructure, and communication systems and thus leads to impoverishment and marginalisation. Bangladesh has limited internal resources available to help cope with catastrophic natural hazards like riverbank erosion and its related consequences. Environmental disasters create the acute problems of unemployment in rural areas, urban slums, and thereby, worsen the socio-economic

[1] Bangladesh consists of a number of administrative areas called division (*bibhag*), each named after its respective capital: Barisal, Chittagong, Dhaka, Khulna, Rajshahi and Sylhet. Every division is further split into 64 districts (*zila*) which are further sub-divided into *upazilas*. The *upazilas* are the lowest level of administrative government in Bangladesh.

condition of the displaced people. Most displacees become landless, which pushes them into further poverty (Zaman, 2007).

Extreme poverty and landlessness forced a significant part of the Bangladeshi population to settle on fluvial islands that emerge periodically. These islands are generally known as *chars* in Bangladesh and are also called *moving islands*.

In Bangladesh, more than 5 million people inhabit the islands of the three major rivers: the Padma (Ganges), Meghna, and Jamuna (Brahmaputra). Consisting of sand and silt, and reshaped by currents during every monsoon, these islands are fertile while also precarious. The melting of the Himalayan glaciers and the subsequent floods often submerge *chars*. When a *char* disappears, it is common that other islands emerge or re-emerge. Although the phenomenon has existed for years, its frequency has increased greatly (Abrar and Azad 2004).

2.2 Cyclones and Tidal Surges, Sea-Level Rise and Salinity

Bangladesh is affected by sudden disasters, such as cyclone Sidr that recently hit the country. Not only the frequency but also the intensity of floods and cyclones has increased. With two extreme weather disasters, the year 2007 was unique in the disaster history of Bangladesh; widespread flooding occurred in July and August, quickly followed by the category-4 cyclone Sidr in November. The flood alone caused 3,363 casualties, affected 10 million people, and reduced crop output by at least 13%. While the flood rehabilitation was underway, the coastal part of the country was hit again by a 240-km/h cyclone, Sidr, that affected 30 districts (out of 64), impacting the lives and livelihoods of 8.7 million people and damaging nearly 1.5 million houses and some 4.1 million trees (ELIAMEP, 2008). Floods and cyclones destroy the crops and cropland. Salinity of the water has increased, causing a shortage of fresh water. Moreover, crops, rice, fish, and trees that need fresh water for survival have decreased. Extreme weather conditions, such as heavy rainfalls and excessive cold, persist even after the main event is over. The lands have lost their fertility. Cultivation of crops without the use of fertilisers is not possible as it used to be in the past.

Over 5 million Bangladeshis live in areas highly vulnerable to cyclones and storm surges. Roughly 55% of the coastal population lives within 100 km of the 710-km long coastal belt of Bangladesh. The majority of those living in this area are low-income agricultural workers; 70% of whom are landless and relatively asset poor (ELIAMEP, 2008).

The country faced 48 major cyclones between 1584 and 2007. In November 1970, between 300,000 and 500,000 people died, and 400,000 houses and 3,500 schools were damaged. During a storm in May 1991, about 140,000 people died, and damage and displacement caused an estimated loss of US $2.4 billion Mirza (1992). By most estimates, the intensity and frequency of cyclones is likely to increase (ibid.). The frequency of cyclones rose from 0.51/year in 1877–1964 to 1.12/year during 1965–1980. The IPCC projected intense and more frequent tropical cyclone activity with extremely high sea level, excluding tsunamis (IPCC, 2007).

If the sea level rises by 1 m, the implications will include a 20.7% land loss, affecting 14.8 million people. The direct and indirect consequences of sea-level rise include saltwater intrusion into surface and groundwater systems, drainage congestion, increased waterlogging potential, and devastating effects on mangroves. About 2.8 million hectares of coastal soil has already become salinised due to heavy withdrawal of surface water and groundwater for irrigation and intrusion of seawater (Zaman, 2007).

3 Migration Patterns

Most experts agree that migration flows in Bangladesh are mainly internal movements from rural to urban areas. Migration occurs principally for economic, social, and environmental reasons. These drivers are not separated from each other but are usually cumulative and mutually enforce each other.

Social networks also play an important role; families tend to remain within their networks when they migrate, and the migration destination choice is often guided by the presence of such networks.

Finally, environmental factors exert a direct and indirect influence on the migration decision and often aggravate economic hardships. Vulnerable people often lose their possessions and means of subsistence during disasters, thus compelling displacement. Although the current impacts of climate change remain disputed, most experts agree that some changes can already be observed, most notably in the seasons' patterns; during the floods and the dry season, people typically undertake a migration to urban areas, in order to increase their means of subsistence.

3.1 Internal Migration

Migration patterns exhibit great differences in the North and the South of the country. To a large extent, these differences can be explained by the difference in environmental changes affecting different regions of the country. In the North, people have developed coping strategies to deal with slow-onset environmental changes, such as erosion or *mongas*, i.e. seasonal food shortages. In the South, environmental changes tend to be more severe and less predictable, and therefore induce more sudden and less-organised migration movements.

Most migrants were interviewed in the slums of Dhaka (Tejgaon, Kawran Bazar, Bangla Motor Uttara, Banani, Mohakhali, Farmegate, and Mohammadpur). All of them reported environmental stress, especially related to the recent devastating cyclone Sidr. Many of them migrated with the hope to recover from the sudden loss. The choice of their destination not only depended on their place of origin and their economic activity and social networks but also on the time when their migration occurred. Following the disaster, households developed different strategies to expand their means of subsistence: in some cases, the fathers migrated temporarily to not only urban centres, such as the capital Dhaka, but also major, closer cities like Khulna or Barisal, in order to work and send remittances. Others migrated in a seasonal fashion and moved where the work was located, and some migrated with the whole family. Finally, some opted not to migrate, since the process is complex and resource-consuming – this financial investment might not be affordable for the poorest families.

In Dhaka, different groups of migrants could be identified: long-term migrants; temporary migrants (either alone or with their family); and men in Dhaka looking for a place for the rest of their family. Generally, women do not migrate on their own.

Migration to Dhaka or India was a forced choice for all migrants interviewed. Life in Dhaka is usually very hard and less satisfactory than expected. Contrary to the official discourse of the *Department of Management of Food and Disasters*, people are too poor to migrate for the sole purpose of social and educational reasons, since these are not their main priority.

Moreover, many slums where migrants had settled have been destroyed by the government, because they were illegal. The government is not implementing any resettlement or rehabilitation plans, but has closed down several slums. In such situations, people move from one slum to the other. At first, these people are determined not to move back to their original places, unless there are possibilities to earn a decent living. However, people are tired of moving all the time without achieving any improvement in quality of life, so when they are faced with this additional and unexpected difficulty (closing of the slums), some migrants chose to return to their place of origin.

Temporary migration is the most common form of disaster-related migration. During the flood season, migration tends to be seasonal: people go to town to find work and send remittances. However, unemployed people and people without assets (such as cattle) tend to move permanently, especially those who lost their land due to riverbank erosion. The loss of land creates unemployment, and less crops can be planted and harvested. If people hope to have better jobs they will most likely stay at the new place: they think that there is no point in going back to the original homeland, as there are no remaining job opportunities.

Factors that influenced the decision to migrate were mainly the unemployment problem, inadequate income, unavailability of croplands, and poor livelihood and living conditions. Environmental degradation was a prime reason for poor income and living conditions. Environmental problems ranged from poor soil conditions and poor water quality to droughts in summer and frequent natural disasters, such as floods and storms. These reasons contributed to the poverty of the people and consequently to their decision to migrate.

The majority of these migrants declared their willingness to move back to their homeland if the environmental situation would improve. Many have left their families behind and want to return back to them. Family networks are strong, and women keep livestock in the villages, which is a good source of income. There is no possibility to keep livestock in the slums. Those who migrated because of bank erosion wish to return to their homelands, with the hope that the rising of new *chars* will bring them new lands. However, people who did not own land are not willing to return to their former places of residence, because they prefer to stay in Dhaka as long as they can work and have a home.

Since the production of crops has decreased due to the intrusion of saline water into the soil, in many parts of the country, growing crops has been replaced by shrimp aquaculture. However, most of the people are not skilled at shrimp aquaculture, and hence, they often end up incurring huge losses. Moreover, shrimp aquaculture is not a labour-intensive activity, which led to unemployment. However, the unemployed cannot return to growing crops due to the deterioration of soil fertility.

People state that environmental degradation is a problem for them at present. As environmental problems worsen, migration tends to be more permanent. According to the Professor M. Elahi[2] interviewed during the field work, it is estimated that about 50% of the people living in slums come from regions where the environment is heavily degraded, and this figure is on the rise.

3.2 Those Who Did Not Migrate

A significant part of the research dealt with people who had not migrated. Most of these people actually had a migration history but have not yet decided to migrate definitely to a more secure area. It has been possible to distinguish between different strategies and behaviours amongst these people:

- People who have started a migration process but have not yet settled in a final destination. These people generally moved to a neighbouring town or village but not yet to an urban area.
- People who initially migrated to India but fell victims of exploitation.
- People unwilling to abandon their assets, such as cattle.
- People who do not have enough financial resources to migrate.
- People resettled by the government or Non-Governmental Organisations (NGOs).
- Women and children, whose presence can be explained either by a family migration strategy, where the husband migrates and sends remittances to his family, or by the fact that most of the victims of major natural disasters were men at sea.

[2] The Professor Dr. K. Maudood Elahi is Pro-vice Chancellor – Head of Environmental Sciences Department of the Stamford University Bangladesh in Dhaka.

In the North, most non-migrants already had a migration history but now move within a very limited area, from one *char* to another, sometimes up to 10 times a year. People consider themselves as non-migrants and develop adaptive strategies to cope with floods and riverbank erosion. They know, however, that they will need to migrate, if the situation becomes worse and prepare themselves to do so. The patterns of limited and localised migration and adjustments to displacement in the floodplain have been conditioned historically by cultural, social, economic, and political factors. A vast majority of displaced families view the river as a symbol for fertility and strongly believe that their lands will re-emerge soon from riverbeds. This does occur periodically, as *chars* for cultivation and resettlement are above normal flood level. Moreover, many poor families cannot afford to move greater distance because of the lack of resources and other uncertainties involved with long-distance migration. Finally, *char* people are a distinctive group, with a distinctive culture. People in most cases do not have title to the land but enjoy a sense of ownership to the property, which prevents them from abandoning their land. The *char* people, especially those living on old *chars*, can enjoy the bounty offered by the silted newly emerged *char* lands. In the upstream region of the Brahmaputra River, land is not as fertile as downstream. The battle for land is therefore less. Moreover, the re-emerged *char* land is most often cultivated by people who do not have entitlement to land or who have a piece of land nearby (Abrar and Azad, 2004). The decision to move from one *char* to another is usually taken by the schools' authorities, and families follow their children when the school is relocated. *Char* people are amongst the poorest in Bangladesh, and the disappearing of *chars* induces conflicts about land tenure, thus aggravating the impoverishment of the population because people lose their title of ownership by losing their land under water and thus become 'tenants' on new land.

3.3 Linkages Between Environmental Degradation and Migration

Even though environmental degradation as such is not yet recognised as a major driver of migration per se, it has intertwining linkages with other migration drivers and greatly affects the job situation of vulnerable people. Furthermore, it leads to problems of exploitation and trafficking.

3.3.1 Impacts on the Employment Situation

After the floods and recent cyclones, families living in coastal regions and along rivers in the South West have lost all their tenure, job opportunities, and sometimes family members. The job situation is shifting both at places of origin and urban areas. For the majority of people living in affected areas, the environmental degradation has made it impossible to earn a decent living. People are forced to find new activities to earn money.

In the North, during the rainy season, the Jamuna river destroys hectares of lands and crops, whereas the same lands are facing desertification during the dry season. It becomes increasingly difficult for the population to make a living out of agriculture, and many men change their activities to fishing. However, this shift has its implications for overfishing. Overall, it is difficult to make a living in the region, and many are compelled to migrate.

In the South, any sea level rise in addition to the lower flow of the Ganges due to a large dam in India results in important floods in the deltaic regions, as seawater is flowing over freshwater from the Ganges. The process is aggravated during the cyclone season. Rice crops are no longer sustainable, and farmers have to adapt. Rice-growing areas have been converted for shrimp cultivation, which has led to many job losses, since shrimp breeding is far less labour intensive than rice crops. Furthermore, the shrimps are too expensive to be consumed locally and the shortage of rice supply drives the price up, leading to food insecurity.

Families develop different coping strategies to overcome these problems, but most of such strategies involve a departure of the man from the household. Some of them start fishing, but fishing becomes increasingly intensive and thus less profitable. Others turn to hunting in the mangroves or the Sundarban islands, but tigers make this activity highly perilous. Finally, some move to large cities to search for jobs. New migrants in cities struggle to find a job and usually become day workers, pulling rickshaws or working in the garment industry.

3.3.2 Women and Human Trafficking

Since most adaptation strategies involve the departure of the men, women are usually left on their own with their children in the villages. Others have lost their husbands at sea during the cyclone Sidr. Women are usually poorer and have less access to resources, which lessens their ability to adapt to environmental change.

Many traffickers take advantage of the situation and smuggle women – and sometimes men – to India and neighbouring areas. Four types of trafficking were observed during the fieldwork:

- Women ended up in prostitution networks or forced labour, as a result of the husband's death or temporary migration.
- Men can also fall victims of human trafficking; after the floods, smugglers promised to bring groups of about 80 men from devastated villages to Northern India, so that they could find work. These men had invested all their savings in the journey but ended up in sweatshops at the Indian border, where they were tortured and treated as slaves.
- Traffickers sometimes targeted entire families.
- Finally, some families resorted to trafficking as a coping strategy, in order to earn a living.

3.4 Role of the Government

The role of the Bangladeshi government in disaster management and early warning systems is weak. The government has not yet fully realised the extent of the threat posed by climate change and has just started implementing a work programme on the issue. A new *Ministry of Food and Disaster Management* has recently been created. This ministry tries to resettle populations in places close to their home areas, even though these places remain at risk. However, no plan exists to deal with population movements in the slums of increasingly saturated Dhaka.

The government has recently taken a more proactive approach to natural disasters: the draft National Policy on Disaster Management (DM) of Bangladesh is a recognition by the Government of the need to shift the approach from disaster response to disaster risk reduction. It also puts much emphasis on involvement of the community and local governments in implementing DM programmes. Nevertheless, a review of Bangladesh flood control strategies suggests that solutions to date have been totally monistic – a single-minded preference for mitigating floods by building dams, dikes, and embankments. The legacy of 'structural' solutions and flood 'prevention' has already led to many detrimental environmental effects of such projects. The narrow view of engineering solutions ignores the more important socio-cultural, economic, demographic, and ecological considerations. 'A "holistic" approach to flood and erosion problems is necessary – one that would recognize the complex interactions of all aspects of how people traditionally have organised, produced, and survived within the physical constraints. Each of these aspects is critical to an understanding of the flood problem from a societal perspective.' (Zaman, 2007).

Some policy recommendations include

- In order to protect people from natural disasters, technical solutions should be taken, such as increasing the level of embankments, elevating the houses, and making culverts to minimise the damage.
- Compensation could be sought from western countries, as they bear most of the responsibility for climate change.
- Free labour movement should be promoted in order to facilitate migration.
- As agriculture is being affected, new agricultural technologies should be promoted. Agricultural supporting industries or small industries in those areas mentioned in this article should be created. Jute industry, fish drying, poultry, and harvesting salt should be introduced as well.
- Solutions for the slums are to encourage the people to go back to their homelands. The Social department of the government is planning to give them land, new incentives, and training.
- Resettlement policies should be developed.

- Insurance systems should be implemented in bank-eroded areas, so that people affected by disasters do not need to wait for foreign help.
- Adaptation and international cooperation in finance and technology should be promoted.
- Finally, education and raising awareness are important. Flood zoning maps and embankments in vulnerable areas can be useful solutions.

4 Conclusions

Flood and erosion disasters are complex phenomena in the current context of Bangladesh. Although uncertainties about the extent of climate change remain, it is a fact that millions of people have recently been affected by floods and '*monga*', requiring evacuation, shelter, and relief assistance.

Floods are a complex mix of both natural and socio-political and economic processes, but the awareness of the impacts of natural hazards on populations in Bangladesh remains modest. Important movements of population are clearly linked to some environmental disasters like riverbank erosion, cyclones, salty water, etc., and climate change will dramatically worsen this situation.

Current movements of population are unmanaged. A similar environmental stress can result in different adaptation strategies and migration patterns. Currently in Bangladesh, migration (whether temporary or permanent) seems one of the best strategies for the affected population to cope with floods, riverbank erosion, cyclones, *monga*, etc. Migration is, however, very complicated and impeded by many barriers; there is no adequate policy or structure of health care for the migrants who are unsure about finding a job or even a safe place to stay during their migration.

People are currently trying to adapt to the first impacts of disasters, but there could come a time when local adaptation will no longer be possible. Bangladesh might then become a theatre of vast population movements, triggered by climate change, which could pose a threat to regional security.

Acknowledgements This research was carried out within the framework of the EACH-FOR Project with the financial support of the European Commission's Sixth Framework Programme (contract number 044468). The project was carried out between January 2007 and March 2009.

References

Abrar, C.R., and S.N. Azad (2004): Coping with displacement, Riverbank erosion in North-West Bangladesh, Dhaka.

Ali, A. (1996): Vulnerability of Bangladesh to climate change and sea level rise through tropical cyclones and storm surges. *Water, Air, & Soil Pollution* 92(1–2), 171–179.

Association for Community Development (ACD). River bank erosion and its impact on displaced people in two districts of Bangladesh, Unpublished working document collected during fieldwork in Bangladesh.

Community-Based Disaster Management Organization (CBDM).(January 2008): Building safer communities in South Asia – Case study Bangladesh, enhancing community solidarity through capacity building and formation of community-based disaster management organizations. 1(1). International Federation of Red Cross and Red Crescent Societies. Available from http://www.adpc.net/v2007/IKM/ONLINE%20DOCUMENTS/downloads/2008/1_CaseStudyCBDMStructureFinal.pdf Accessed on 17 March 2009.

ELIAMEP. (2008): Gender, climate change and human security, lessons from Bangladesh, Ghana and Senegal, ELIAMEP, Athens.

Hossain, M.M. (1991): Shifting characteristics of the Ganges in Bangladesh. In: Elahi, K.M. Ahmed K.S., and Mafizuddin, M. (eds.), *River Bank Erosion Impact Study*. Jahangirnagar University, Savar.

IPCC (Intergovernmental Panel on Climate Change). (2007): Summary for Policymakers. Climate Change 2007: Impacts, Adaptation and Vulnerability. Contribution of Working Group II to the Fourth Assessment Report of the IPCC: 7–22.

Mahtab, F.U., and B. Karim. (1992): Population and agriculture land use: Towards a sustainable food production system in Bangladesh. *AMBIO: A Journal of the Human Environment* XXI(I), 50–55.

Mizra, M.M.Q. (1992): *Natural Disaster and Environment in Bangladesh*. Centre for Environmental Studies and Research, Dhaka.

Zaman, M. (2007): Flood and erosion disasters in Bangladesh: The human dimensions, Vancouver.

Chapter 17
In the Land of Ostriches: Developmentalism, Environmental Degradation, and Forced Migration in Turkey

Zeynep Kadirbeyoglu

1 Introduction

This chapter examines the impact of environmental change on forced migration in the case of Turkey. Environmental change includes floods, air, soil, and water pollution, and large development projects such as dam construction. These types of changes in the environment of a region may cause massive disruption and force people to migrate. In some cases, such as earthquakes, migration may be temporary, while in other cases, such as those displaced by flooding of a dam reservoir they move permanently and sometimes are resettled in distant areas with different cultural and social environments. Turkey is likely to face severe environmental problems in the coming decades as a result of climate change. However, migration as a result of environmental degradation has not really been a major concern for politicians, academics, or NGOs. This chapter fills this gap by examining two cases: the construction of the Ataturk Dam in the southeast of Turkey and the depletion of groundwater sources in Suruç, both of which led to significant migration from the respective localities.

The main findings of this chapter are that in the case of displacement caused by a development project, the state's role is imperative in mitigating the negative social and economic impacts. Consultation with potential migrants prior to project implementation is important for the determination of where to resettle them, if they wish to be relocated by the state. The employment opportunities in the new area of residence are important to facilitate the transition process. Especially important is giving small landowners and landless farmers the option to be resettled by the state where they could get a small plot of land and a house. This requires significant research by the state and considerable financial resources for implementation. Hence, reporting the construction costs of development projects could be misleading in terms of total costs, because they ignore the social and environmental costs.

Z. Kadirbeyoglu (✉)
Bogazici University, Istanbul, Turkey
e-mail: kadirbey@yahoo.com

T. Afifi, J. Jäger (eds.), *Environment, Forced Migration and Social Vulnerability*,
DOI 10.1007/978-3-642-12416-7_17,

In the case of drought-induced forced migration, the most important finding is the lack of state involvement in finding a solution to the gradual environmental degradation that took place since the 1980s. The drought affected not only the farmers whose main source of livelihood – cotton and wheat production – disappeared but also other sectors – services, transportation, and construction. The villagers migrated to the district centre, Urfa city centre or Istanbul, Izmir, and Mersin. Those who could not find employment to bring their families along resorted to seasonal migration for agricultural employment or other services, such as tourism and construction. Most of the interviewees indicated their desire to move back to Suruç, if water were provided by the pending irrigation project of Southeast Anatolian Project.

2 Theoretical Framework

Environmental degradation including climate change has the potential of affecting the lives of millions in the coming decades. Environmental change has already led many local communities to migrate within their own state borders or attempt to move to other countries for survival. Within this context, academics and policymakers are trying to bridge the gap between these two heavily researched fields – migration and environment – which have not really interacted to the extent that would be desirable given the potential adverse effects of the issues. Suhrke (1993: 9) defines environmental migrants as those 'who respond to a combination of push-pull factors, environment being only one of them'.

Despite rising concern about the environment–migration nexus, especially in the case of Turkey, there is little empirical work that seeks to determine the impact of environmental change on migration. In the political science literature, internal migration has been associated with modernisation up until the 1960s. The observed trend and theoretical forecast was that rural–urban migration increases with industrialisation and that the ratio of urban dwellers rises as the number of those working in agriculture declines. This has happened in the case of Turkey and this migration process attracted substantive academic interest. In Turkey, economic development, which started gaining pace in the 1950s, increased already-existing regional imbalances in incomes and living standards, such that migration became attractive for many rural inhabitants who dreamed of a better future. Other factors that accelerated the migration process were increasing population levels and mechanisation in agricultural production (Ozcan, 1998: 78). The continuous increase in urbanisation rates meant that while 19% of the population was living in urban areas in the 1950s, this number had increased to 65% in 2000 (DPT, 2003). Icduygu and Unalan (1998: 43) claim that approximately half of the increase in urban population was due to rural–urban migration that took place since the 1950s.

Forced migration has been a constant reality in Turkey. Historically, the main sources of forced migration have been political and social conflict, large-scale development projects, and natural disasters. Earthquakes are the type of natural disaster that cause the most abrupt and widespread disruption. However, migration that results from earthquakes has tended to be short term.[1]

Unlike development-induced displacement, water scarcity, air pollution, and declining yields tend to occur more gradually but can have more significant impact on migration. Whereas for the development-induced displacement the affected receive aid and have the option of being resettled by the state, slowly progressing environmental degradation tends to attract less attention and a less brisk response from authorities and civil society organisations.

Studies that discuss the problem of forced migration and refugees generally focus on the causes of migration and the status of these people in international law (Turton, 2003: 17). However, this narrow perspective is unable to capture the individual experiences of migrants during the process of decision-making, migration, and resettlement. Following Turton's (2003: 17) suggestion that studies should focus on the experience of forced resettlers/forced migrants/ refugees, this chapter examines the experience of people who moved due to environmental problems and development projects.

3 Synthesis of Context

In Turkey, state-led development focused on industrialisation of the economy in line with modernisation attempts, without much regard for the environment or the conservation of natural resources (Adaman and Arsel, forthcoming). This developmentalist approach to economic growth was not mitigated by an organised civil society which could buffer the negative impact of modernisation-oriented development on the environment. This is one of the reasons why there was scant interest in environmentally induced forced migration: policy-makers saw it as a natural/normal phenomenon and ignored it by burying their heads in the ground just like ostriches. A sacrifice for the development of the country was seen as the duty of all citizens.

Most significant environmental problems in Turkey stem from (a) rapid and unplanned urbanisation and industrialisation, which led to air and water pollution; (b) intensive agricultural production – excessive use of pesticides and insecticides, which led to water and soil pollution; (c) erosion; (d) toxic wastes; and (e) marine pollution (Koymen, 1999; TCCOB, 2004: 196; Okumus, 2002; Ecevit and Ecevit, 2002). Starting from the early 1980s, the increasing number of tourists caused environmental degradation due to the consequent increasing demand for water, which in turn reduced the groundwater reservoir levels (TCCOB, 2004: 152). These are all important problems but they remain outside the scope of this chapter as the focus is on the migration caused by dam building and groundwater depletion. Furthermore, the assumption that people will migrate out of areas with significant pollution (air, water, soil) does not hold

[1] See, for instance, Palamut, 2007 for the case of the Marmara earthquake of 1999.

in Turkey. On the contrary, the industrialised and polluted areas attract individuals who are pursuing economic improvement.

The first Environment Code was legislated in 1982 and was followed by various regulations focusing on air, water, and soil quality, disposal of hazardous waste, etc (Okumus, 2002). This code, however, was not effectively implemented. Adaman, Ozkaynak, and Hakyemez (forthcoming) argue that this ineffectiveness is due to three factors: first, the state adopted many of the environmental codes as a way of responding to global institutions; second, the prevalence of patron–client relations within a top–down modernisation project eroded the public sphere; and third, the institutions that were supposed to regulate the environment were designed with counterproductive incentive structures. In the post-1980 era, the state was unwilling to implement the newly emerging environmental regulations, despite increasing manifestations of environmental activism in Turkey (for example, see Adaman and Arsel (2005)).

4 Methods

4.1 Justification of Case Selection

The link between environmental degradation and migration has not been explicitly examined in Turkey. The multiple causes of migration are acknowledged by all prominent migration scholars in the country but systematic research is still missing.

The Ataturk Dam – for which the construction began in 1983 and was completed in 1992 – on the Euphrates River in the southeast of Turkey is the largest dam in Turkey and is part of the Southeast Anatolia Development Project (Guneydogu Anadolu Projesi – GAP) (Kibaroglu, 2002; Morvaridi, 1990; Tortajada, 2000). The case of the Ataturk Dam is examined in order to evaluate the experiences of the people displaced by this project. The second case that is analysed is the case of Suruç, where groundwater reservoirs were depleted in the 1990s and the majority of the population was forced to migrate. Suruç had abundant groundwater and farmers began cultivating cotton by using pumps – pumps were installed starting from the late 1950s. Over the years, the level of groundwater started to drop significantly. Farmers had to dig deeper wells and use more powerful pumps in the 1970s. The water started to vanish in the 1990s. Those whose livelihoods depended on irrigated agriculture migrated to the city centres in the region or to other cities throughout Turkey. Some became seasonal migrants and rented land in other districts or provinces or became sharecroppers and continued agricultural production, while others moved to other sectors. Those who stayed behind in Suruç are mostly the elderly population and those who could not afford to migrate.

4.2 *Methodology*

The fieldwork for the case of Turkey began in September 2007. I carried out 14 expert interviews with academics, NGOs, ministry officials, and migrant organisations; 28 interviews with people who were displaced by the Ataturk Dam; and 20 interviews with those who had to leave Suruç due to the depletion of groundwater and drought. Fieldwork in Urfa and Suruç was carried out in October/November 2007. The fieldwork for the dam-induced forced migrants to Izmir–Torbali and Izmir–Bulgurca and drought-induced migration to Istanbul was carried out in March/April 2008. The access to those who migrated from Suruç because of drought was more difficult than reaching those who self-settled in Urfa. Many of them lived in Adana, Izmir, Istanbul, and Ankara. I focused on those who had migrated to Urfa and Istanbul city centres. Snowball sampling was used in the case of Suruç migrants as well.

5 Development-Induced Forced Migration: Ataturk Dam

The Ataturk Dam flooded one district entirely (Samsat) and affected 85 settlements from Adiyaman, 48 from Urfa, and 12 from Diyarbakir; 113,476 people from 18,121 families were impacted by the construction (Guler Parlak, 2007: 11). The inhabitants of these villages were given two basic options: either to get the confiscation compensation and self-relocate, or to be relocated by the state either to the urban area within the region or to a rural area outside of the region. The case of those relocated by the state was studied by Guler Parlak (2006, 2007) and the results from her studies are presented here together with the findings of the fieldwork.

Guler Parlak (2006) examines the case of relocation to New Samsat, whereas Guler Parlak (2007) reports the findings from research with groups who were relocated to a village in the west of Turkey. Research was carried out with three groups: those who decided to self-relocate to Urfa city centre; those who chose to self-relocate to Torbali–Izmir; and those who chose to be resettled by the state to Bulgurca–Izmir in the west of Turkey. Previous studies had not contacted the self-settled families, as they are hard to track. However, it is important to examine this group, since most studies of displaced people found that those who receive the confiscation compensation tend to misuse the money – not having previously come across such amounts at once – and face significant deterioration in their living standards.

The goal of the chapter is to evaluate the adaptation of the displacees to their new circumstances, their economic and social well-being, and the changes in their livelihoods. Preliminary findings indicate that migration was not an easy process and that there were significant hardships.

5.1 Settling in the West of Turkey: Yalikoy and Bulgurca

This section is based on the study of Guler Parlak (2007) on migrants to Yalikoy–Didim and my fieldwork carried out in Bulgurca–Izmir. To accommodate some of the displacees, a new village was founded on state-owned land in Aydin: Yalikoy. Migrants to Yalikoy were given houses and 30–40 *dönüm*[2] of land. The land that was allocated to these displacees was previously used by the locals, and this created tensions between those who were resettled and the local inhabitants surrounding their village. Those who were resettled in Bulgurca were given housing that was originally constructed to resettle families from the Izmir region.

The most important problem for the displacees was the waiting period between the flooding and the move to the new settlement. People had to wait between 2 and 10 years prior to being resettled by the state. Many families were split, since it was impossible to resettle entire extended families together in the new locations outside the region. Unemployment and discrimination were the most important problems that displacees faced once they were resettled. While the younger generations appreciate living in the west of Turkey, the elderly want to go back to the southeast.

5.2 Settling Near the Old District: New Samsat

Samsat had 33,800 ha of land prior to the flooding of the dam reservoir. Including the district centre, 19,341 ha were submerged by the dam's waters (Guler Parlak, 2006: 53). New Samsat has 550 households in the district centre, 209 of which received public housing provided by the state. Following displacement, 773 households were given land to construct their own dwelling, but only 250 of these families built a house. The rest probably migrated to Adiyaman city centre or other cities (Guler Parlak, 2006: 11).

Resettlement in Samsat created some problems. There were major problems with the confiscation compensation payments for those who chose to self-relocate. Many had misrepresented the value of their old home, thinking that the state was registering their holdings for taxation purposes. These families ended up getting much less compensation than the value of their real estate. Most of the large landowners – who did not have such valuation problems by virtue of being more informed – moved to city centres and built a new life, making use of their confiscation compensation. Small landowners and landless people were resettled in New Samsat.

Many families were producing their own foodstuff prior to the dam, but they had no space for that in New Samsat (Guler Parlak, 2006: 94). The most important problem of migrants living in New Samsat is unemployment.

---

[2] A dönüm is approximately equal to 1 decare (approximately 10 *dönüms* is one hectare).

Residents of New Samsat – often the whole extended family including the children and the elderly – have to travel to different parts of Turkey to work as seasonal agricultural labourers. In contrast to the case of Yalikoy, where 31% of the inhabitants claimed that their economic situation had worsened following the displacement, 72% claimed this to be the case in Samsat. As part of the negative impacts of the dam construction, out-migration from Samsat increased following the flooding of the dam reservoir. According to the survey Guler Parlak (2006: 102) conducted with 200 households in New Samsat, families with out-migrants increased from 21 to 57%, following the dam construction. A significant portion of those who wanted to migrate from New Samsat wanted to go to city centres outside their region (38%) while 34% wanted to settle in city centres within their region.

5.3 Self-Resettlers: Urfa City Centre and Torbali–Izmir

Among those who chose to self-relocate from the villages that were being submerged by the dam reservoir, many chose to move to Urfa or Adiyaman city centres. The disbelief that the dam would flood their village was the most important problem. Those who resettled in Urfa suffered from high real estate prices and unemployment. Many migrants work as seasonal agricultural workers. Some, however, moved to villages in the west of Turkey, because they had relatives who had migrated there before. The group that moved to Torbali–Izmir on its own chose to receive the confiscation compensation, because they did not believe that the state would provide appropriate housing. In most cases, the whole family now lives in Torbali – unlike the case of the split families of Yalikoy–Didim. Most migrants to Torbali work as hired farmers, whereas in the past they were all landowners. Most migrants claim that the dam made their situation worse but only a few of them said that they would move; they either felt comfortable with the utilities they had access to at home or they were too old to migrate anywhere.

Villagers who were displaced were not properly informed about the dam construction. The headmen and *aghas* (large landowners) were invited to the information meeting. They reported to the villagers that a dam would be built and that they would have to leave their homes. People found it hard to believe, even when State Hydraulic Works[3] officials visited their village and told them informally that a dam would be built. This created great confusion and disbelief among those whose homes would be flooded by the dam reservoir. They were caught by surprise when the waters started to rise and swallow their belongings and homes. They escaped to higher ground and lived in tents for some time before moving to the cities or being resettled (Guler Parlak, 2006: 119). Among

---

[3] State Hydraulic Works is the state agency which plans, builds, and operates dams in Turkey. The Turkish name for the agency is *Devlet Su Isleri* (DSI).

those who were resettled in New Samsat, 78% claimed that they had not been informed prior to migration (Guler Parlak, 2006: 117). Lack of guidance prior to displacement continued afterwards: 98% of the families in New Samsat claimed that they had not received any information or guidance after they moved (Guler Parlak, 2006: 117).

All migrants referred to the insufficiency of compensation amounts, especially given the family size (average household size is seven in the region). They had never received this much money and it mostly went on expenditures and vanished in no time. Some migrants who received the confiscation compensation invested the money in businesses which then failed. An old man who used to own large stretches of land wasted the compensation he received and had to go back to the village – which has been moved up the hill by the people who fled the reservoir. He is now making a living as a fisherman.

The confiscation compensation was not properly invested in productive enterprises. Guler Parlak (2006: 140) claims that there are several reasons behind the gradual loss of this capital that could have been used more productively. The first one is the unavailability of accurate information or guidance. Second, most of these peasants received a large sum of money for the first time in their lives and mistakenly thought that this money would never end. Third, many believed that receiving interest on the money was a sin and just kept the money. Fourth, they received the compensation gradually as water started to flood and they could not make good use of money that came in instalments. Fifth, many had no other job experience than being agricultural producers and were unable to succeed in other jobs.

Social and economic vulnerability increased the likelihood of suffering from the displacement. Large landowners were able to get large sums of confiscation compensation and were able to maintain their living standards as long as they were able to invest it. In contrast, small holders either asked to be resettled by the state and suffered from separation from their families, the difficulties of migration, and adaptation to new circumstances, or they got their confiscation compensation and tried to establish a new life with limited financial and social capital. The most hard-hit were the landless peasants who were not given any compensation, unless the headmen registered them as part of the groups which would be resettled by the state.

There were problems of adjustment to the new environment. The unwelcoming and sometimes even racist attitude of the non-Kurdish population of the relocation destination caused problems. The fact that they were given land was resented by the villagers who live in the area where they were resettled. Guler Parlak (2007) argues that the problems of adaptation could be reduced, if the state attempts to resettle villagers to state-owned land close to their villages. Her survey results indicate that 48% of the settlers claimed their main problem to be adaptation problems to their new environment. The second most common problem among the settlers was unemployment and hardship, especially when they first arrived (Guler Parlak, 2007: 156). Fifty-two percent of those who were resettled claimed that they sometimes think it would be better, if the dam had not been constructed. The most distressing part of the resettlement seems to be the separation from relatives and neighbours (Guler Parlak, 2007: 158).

6 Drought-Induced Forced Migration: The Case of Suruç

Suruç is a district of Urfa and has a population of 82,247 according to the 2000 population census. There are 82 villages, 134 hamlets, 1 town, and 1 district within Suruç. Previously Suruç was predominantly an agricultural district. Cotton, wheat, barley, and lentils were the main crops of the district. Currently, only a minority of Suruç farmers have access to irrigated agriculture. Abundant groundwater levels – which used to spring from 1 m below the surface – began falling in the 1970s. Farmers who were warned by some of the state officials disregarded the warnings and kept increasing the depth of the water wells. This fight over water had to stop when water levels were so low (approximately 200 m) that it did not make economic sense to dig even more to get water. Abundant groundwater was depleted in a matter of 25–30 years.

Winter rainfall, which is of crucial importance for the agricultural production, has declined by 60% over the past decade as well. In March 2008, rainfall in the Euphrates Basin – where Suruç is located – was 54% lower than the normal level of March rainfall, according to the Directorate of Meteorology statistics. Nowadays, drinking water is scarce in the villages of Suruç, let alone water for irrigation. Water tankers carry water to the district to provide the inhabitants with daily supply. These conditions of drought caused some people to leave Suruç permanently and migrate to other cities, whereas others leave as seasonal agricultural workers during the summers.

In most cases, wealthy families were able to migrate to city centres, such as Urfa, Ankara, or Istanbul. The less well-off still reside in the villages and maintain their livelihood by working as seasonal agricultural workers in Adana, Aydin, and Harran. Most of these migrants leave in the summer and return to Suruç in the winter. A headman from a village on the Syrian border claimed that his village and its 3 hamlets had 30–40 households in the past. As a result of migration the number of households decreased to 10. In the past, even those who had small plots could make a living, because inputs were not as expensive as today; there was abundant water and cheap credit. The headman states that these remaining households rent land in Harran or other villages of Urfa, where there is irrigation (Suruç Interview #5).

Among the 20 interviewees from Suruç, 2 still reside in the district. The others either live in Urfa city centre or in Istanbul. Only one migrant stated that he was going back and forth between Suruç and Istanbul. While 11 migrants were originally from Suruç district centre, 9 of them came from various villages of Suruç. All migrants (except one who migrated following marriage) stated that the water shortage and drought, which caused unemployment for agricultural producers, merchants, truck drivers, and shop and restaurant owners, were the main reasons for their migration.

Except for the temporary migrant who goes back and forth and makes a living as a baker in Istanbul – where he gets paid three times what he would get in Suruç –, families of the interviewees had migrated out of Suruç. Some migrants had family members in Germany or Russia. However, many of these migrants did not send back remittances and did not support their families back in Turkey. Migrants from Suruç, migrated to destinations where their relatives had already migrated – Istanbul, Ankara, Mersin, and other big cities. Many of them were supported by their relatives in the process.

In brief, the major cause for migration from Suruç is the depletion of groundwater and drought, which caused ripple effects in all sectors of the economy. The tension in the political sphere is also a significant cause pushing households to migrate. Most migrants were helped by their relatives who had already migrated out of Suruç. Many face discrimination in Istanbul and would like to go back, if Suruç would get water from the dam reservoir. The most important finding of this case is that the lack of state regulation of wells and water consumption can be detrimental in the long run, even if in the short-run farmers prosper thanks to irrigated agriculture.

7 Conclusion

This chapter examined the cases of dam-induced and drought-induced forced migration in order to understand the role of environmental factors in inducing migration. In the case of the Ataturk Dam, migrants to the west experienced problems of adaptation and employment. Adaptation to a very different social environment was more difficult for the elderly. Despite difficulties, the majority of migrants are satisfied with their houses and are happy to be in Yalikoy. Those who resettled in New Samsat had no problems with adaptation but had many difficulties in finding a job and maintaining their livelihood following the flooding. The hardest hit from the displacement resulting from dam construction were those who chose to self-resettle. There are indications that some of these people, who received the confiscation compensation in three instalments, used up the money on consumer goods and were unable to buy a new house. Not only the inability of displacees to manage their funds but also the sky-rocketing real estate prices in cities like Urfa and Adiyaman made resettlement difficult. There is a distinction between large landowners and small ones: large-landed families received a considerable amount of compensation and were, for the most part, able to establish a new life and acquire a sustainable livelihood in the post-relocation process. Those who did not own land and did not sign up for the state resettlement were the hardest hit by the process – as the example of some of the migrants to Torbali–Izmir indicates.

Forced migration in Suruç is mostly a result of environmental degradation and scarcity of water resources. The case of Suruç is a much clearer example of environmental degradation and forced migration. The depletion of groundwater levels in Suruç led many families to migrate. They were agricultural producers prior to the drought in Suruç. Most plots are still cultivated in Suruç by those who are left behind, but without water they do not get much yield, especially with declining rainfall in the last couple of years. This is why many of the remaining people in Suruç work as seasonal agricultural workers or sharecroppers in other regions. They are usually seasonal migrants and do not make much money. Those who stayed behind in the villages of Suruç are the ones who were unable to find the means to migrate to district centres or other large cities. Thus, economic and social vulnerability forces those who are disadvantaged to remain as such. Seasonal migration and sharecropping in other districts/provinces prevent younger children – who work in the fields and accompany their parents in the agricultural season – from attending school for the first 2 months of the school term. In contrast to the case of Ataturk Dam, state officials provided no support for those who faced difficulties in Suruç.

Most forced migrants from Suruç face economic and social difficulties in their new area of residence. It is hard for those who were farmers in the villages of Suruç to find employment in the cities, and their incomes cannot maintain their livelihoods, since they have to purchase foodstuff which they used to cultivate back home. They cannot provide the essential staples or any help from the villages either – which is how migrants survive in larger cities until they adapt to the job market – because of drought. In addition to economic difficulties, those who migrated to larger cities were confronted with discrimination and racism because of their Kurdish identity. They look back to old days in Suruç with longing and claim that they would go back, if Suruç were to get water from the Ataturk Dam. Only a few claimed that they would not go back, because they could not re-adapt to life in Suruç.

Acknowledgements The author would like to thank Muge Aknur and Efe Baysal for their help in the research process. The usual caveat applies. This research was carried out within the framework of the EACH-FOR Project with the financial support of the European Commission's Sixth Framework Programme (contract number 044468).

References

Adaman, F., and M. Arsel, (eds.) (2005): *Environmentalism in Turkey: Between Democracy and Development*? Ashgate, Aldershot.

Adaman, F., and M. Arsel. (Forthcoming): Globalization, Development and Environmental Policies in Turkey. In: Cetin, T., and Yilmaz, F. (eds.) *Understanding the Process of Institutional Change. In Turkey: A Political Economy Approach*, New York: Nova.

Adaman, F., B. Ozkaynak, and S. Hakyemez (2009): The political ecology of a Ramsar site conservation failure: The case of Burdur Lake, Turkey. *Environment and Planning C* 27(5): 783–800.

DPT (Devlet Planlama Teskilati-State Planning Institute). (2003): *Sosyo-ekonomik gelismislik siralamasi (Socio-Economic Development Ranking)*. DPT, Ankara.

Ecevit, M., and Y. Ecevit. (2002): Kirsal Yosullukla Mucadele: Tarimda Mulksuzlesme ve Aile Emeginin Metalasmasi (Struggling with rural poverty). In: Ozdek, Y. (ed.), *Yoksulluk, Siddet ve Insan Haklari (Poverty, Violence and Human Rights)*. TODAIE, Ankara.

Guler Parlak, Z. (2006): *Suya Batan Gunes: Samsat (Sun Setting in Water: Samsat)*. Turhan Kitabevi, Ankara.

Guler Parlak, Z. (2007): *Yasamin Suyla Dansi: Barajlar ve Surdurulebilir Kalkinma (Dance of Life with Water: Dams and Sustainable Development)*. Turhan Kitabevi, Ankara.

Icduygu, A., and T. Unalan. (1998): Turkiye'de Icgoc: Sorunsal Alanlari ve Arastirma Yontemleri (Internal migration in Turkey: Problem areas and research methods). In: *Türkiye'de İç Göç,Türkiye Ekonomik ve Toplumsal Tarih Vakfi*, İstanbul.

Kibaroglu, A. (2002): Design and management of irrigation systems: The Southeastern Anatolia development project. In: Ozay, M., and Bıcak, H.A. (eds.), *Modern and Traditional Irrigation Technologies in the Eastern Mediterranean*. International Development Research Centre, Ottawa.

Koymen, O. (1999): Cumhuriyet Doneminde Tarimsal Yapi ve Tarim Politikalari (Agricultural structure and policies during the republic). In: Baydar, O. (ed.), *75 Yilda Koylerden Sehirlere (From Villages to Cities in 75 Years)*. Tarih Vakfi Yayinlari, Istanbul.

Morvaridi, B. (1990): Agrarian reform and land use policy in Turkey: Implications for the Southeast Anatolia project. *Land Use Policy* 7(4), 303–313.

Okumus, K. (2002): *Turkey's Environment: A Review and Evaluation of Turkey's Environment and Its Stakeholders*. The Regional Environmental Center for Central and Eastern Europe, Szentendre.

Ozcan, Y.Z. (1998): Icgocun Tanimi ve Verileri ile Ilgili Bazi Sorunlar. In: Türkiye'de İçgöç. İstanbul, Turkiye Ekonomik ve Toplumsal ve Tarih Vakfi, Istanbul.

Palamut, H. (2007): Depremin Yarattigi Travmanin Kisilerin Hayatlarina Etkileri: Golcuk Ornegi. In: Kasapoglu, A. (ed.), *Yeni Toplumsal Travmalar: Gunumuz Turkiye'sinde Oykulesen Magduriyetler (New Social Traumas: Stories of Grievance from Today's Turkey)*. Referans Yayinevi, Ankara.

Suhrke, A. (1993): Environmental degradation, migration and conflict. Paper for the workshop on Environmental Change, Population Displacement, and Acute Conflict, Institute for Research on Public Policy, Ottawa, June 1991.

TCCOB. (2004): *Turkiye Cevre Atlasi (Turkey's Environment Atlas)*. Turkiye Cumhuriyeti Cevre ve Orman Bakanligi, Ankara.

Tortajada, C. (2000): Evaluation of actual impacts of the Ataturk dam. *Water Resources Development* 16(4), 453–464.

Turton, D. (2003): Refugees, forced resettlers and 'other forced migrants': Towards a unitary study of forced migration. UNHCR, *Working Paper No. 94*, Geneva.

Chapter 18
Environmental Migration: Case of Kyrgyzstan

Emil Nasritdinov, Mehrigul Ablezova, Jypara Abarikova, and Aigoul Abdoubaetova

1 The Environmental Situation in Kyrgyzstan

The Kyrgyz Republic covers an area of 198.8 thousand km^2 (5.3% – forests, 4.4% – lakes and rivers, 54.1% – agricultural lands, and 36.2% – other) and about 96% of its territory is located more than 1,000 m above sea level. About 30% of the territory of the Kyrgyz Republic is habitable, whereas 70% of the territory are high mountains. The Pamir–Altay mountain range is in South West. In the North East, there is the Tien Shan with the region's highest point – Pobeda Peak (7,439 m). These mountain ranges are divided by great valleys and gorges; in the North – the Chui and Talas valleys, in the South West the Fergana valley, and in the South, the Alay valley. Both the Pamir and Tien Shan Mountains are still young and seismically very active (Stepanenko, 1997). This seismic activity and the 7,000 m difference between high and low altitudes are among the main reasons for frequent natural disasters, the larger share of which occur in Southern Kyrgyzstan (MES, 2008).

1.1 Natural Disasters

The territory of the Kyrgyz Republic is considered geologically to be the most dangerous area of Central Asia, and it is subject to more than 20 kinds of dangerous natural processes that cause natural disasters like earthquakes, landslides, floods, stone falls, avalanches, etc. The distribution of the frequency of the major emergency situations in the Kyrgyz Republic is the following: floods – 30%, landslides – 16%, technogenic – 10%, earthquakes – 9%, and meteorological – 9% (MES, 2008)

Annually, seismologists record about 3,000 earthquakes, tens of them are strong, although as a rule, only a few cause damage to houses and buildings. According to geological specialists, more than 200 localities are categorised as seismically dangerous. In 2006, there was an earthquake of magnitude 7 on the

E. Nasritdinov (✉)
Social Research Center, American University of Central Asia, Bishkek, Kyrgyzstan
e-mail: emilzn@gmail.com

T. Afifi, J. Jäger (eds.), *Environment, Forced Migration and Social Vulnerability*, DOI 10.1007/978-3-642-12416-7_18, © Springer-Verlag Berlin Heidelberg 2010

Richter Scale (RS) in the Naryn oblast. Its epicentre was in Kochkor village, and this caused damage to 7,003 houses and 50 public buildings; 7 were completely destroyed and 1,322 were severely damaged. On 8 January 2007, there was an even larger earthquake in Batken, scoring 8 points on the RS. It damaged 7,137 houses, 90 of them severely (MES, 2008).

More than a half of the territory of the Republic is subject to avalanche danger. The duration of the avalanche season is 5–7 months. Every year avalanches cause frequent emergency situations on the mountain roads. More than 10 river basins and 800 avalanche-endangered areas have been identified in the country; 50 people died under avalanches during the period of 2002–2007.

Because of intensive glacier thawing as well as plentiful snowfall and rainfall, many regions of the country are subject to flooding every spring. There are 3,100 river flood basins in the territory of Kyrgyzstan. In addition, 200 of the 2,000 Alpine lakes have the potential for causing floods. In 2003, 38 locals of the Sogot village in the Uzgen region died from flooding. In 2004, in the area of the Kaynama Budalyk village (Alay region), a flood killed 33 people, 17 of whom were children. Furthermore, in April 2004, in the Karasogot area of the Karasu region of Osh oblast, a flood with a volume of 60,000 m^3 took the lives of 5 school children. At the end of March 2005, a flood on the territory of the Sarybulak village (Karakuldja region) destroyed the road and wiped out 700 m of the electrical lines and 1.5 km of the communication lines. In 2003, after the flood, 50 families from this village were relocated.

Landslides are another major and common environmental disaster in Kyrgyzstan. Eighty-eight people were killed by them during the period from 2002 to 2007 (MES, 2008). Currently, there are more than 5,000 active landslides in Kyrgyzstan. More than 509 settlements are – or can be – affected, and more than 10,000 families live in endangered houses. Floods and landslides can also be seen as the result of human activity. The main reason for the increase in their numbers in recent years is soil erosion on mountainous slopes, which is caused by the overgrazing of pastures. Later in the chapter, this is analysed in more detail. In the next section, the chapter addresses a more explicit anthropogenic activity affecting environment – uranium mines.

1.2 Anthropogenic Disasters

The extraction of uranium, heavy metals, and mercury, along with the accumulation of their wastes, is among the main human-made environmental disasters in Kyrgyzstan. They cause environmental pollution and constitute an already existing and potential health danger. These threats are aggravated by the fact that the Kyrgyz Republic is situated at the upper portion of the region's water basins. Therefore, contaminated substances from its territory can reach the other regions of Central Asia through numerous water flows. There are five main uranium tailing-fields in Kyrgyzstan: Mailuu-Suu, Kara-Balta, Ming-Kush, Kaji-Say, and Ak-Tuz.

The extraction of uranium in Kyrgyzstan started in the 1940 s as a part of the Soviet nuclear programme. It was aimed to provide nuclear power stations in Russia with raw materials. At that time, there was restricted access to the Kyrgyz towns and villages situated near the uranium pits, since the former were classified as 'secret areas'. The extraction of uranium in Kyrgyzstan continued until 1968. Afterwards, the pits were closed, but the tailings, where the uranium wastes were stored, remained. There are more than 250 million m^3 of wastes in 92 locations where the mining industry existed on the territory of the republic; they all contain radionuclides which are hazardous and toxic for humans. After the collapse of the mining corporations of the Soviet Union, these 36 tailing-fields and 25 pits with a total volume of 15.7 million m^3 were placed under the authority of the Ministry of the Emergency Situations (MES) of Kyrgyzstan. They included

- Thirty-one tailing-fields with radioactive wastes – 7.2 million m^3;
- Five tailing-fields with toxic wastes – 5.2 million m^3; and
- Twenty-five mountain pits of un-recovered ores – 3.3 million m^3.

Potentially dangerous natural processes such as landslides and floods were not taken into account during the planning and construction stages of these sites. Most of the tailing-fields are situated in river-beds, which continuously erode them, resulting in discharges of radioactive substances into the environment: soil, atmosphere, and subsoil waters. The situation is complicated by the activation of landslides and earthquakes in the tailing-fields zone. Scientists state that there is a real danger of the discharges of radioactive wastes into Kyrgyzstan's rivers. In the case of possible pollution, the water arteries of Kyrgyzstan belonging to the basins of Syrdarya and Amudarya Rivers can quickly spread radioactive and toxic wastes into the territory of Uzbekistan and Tajikistan. A large-scale discharge would cause a regional catastrophe and threaten the existence of a large part of the local population and environment. Currently, the MES works on reburying the wastes. From 1999 until 2007 it spent 8.8 million som, but only temporary safe conditions of tailing-fields were provided (Gruzdov, 2003; Mamajakyp uulu et al., 2005; Radjabov, 2008; Zvyagelskaya, 2001)

2 Migration

Almost half of the Kyrgyz population lives in poverty (55% of the rural and 28.3% of the urban population); 16.9% of the rural and 6.9% of urban population live in extreme poverty (Tishin, 2007). As a result of economic and political reforms in the 1990s, migration became a top priority issue in the Kyrgyz Republic. High unemployment rates, the reduction of living standards, and the lack of social protection are the main factors causing large-scale spontaneous migrations, both internal and external. It appeared that migration peaked in the period between 1991 and 1993. Around the year 2000, migration processes became more stable and showed lower rates, but after the events of 24 March 2005, these processes increased again. About 400,000 citizens of the Kyrgyz Republic leave the country every year. Most of them travel to Russia and Kazakhstan as labour migrants. This is connected to a high demand for labour there, better salaries in these countries, and extreme poverty, especially in Southern Kyrgyzstan.

This scale of migration aggravates the demographic problems in the Republic, which is characterised by the decreased birth rate and increased death rates – thus, a decrease in the population growth of an already low-density population country. The intensity of migration processes and decrease in the natural growth negatively affect the labour potential of Kyrgyzstan and strengthen a tendency towards an 'ageing' population.

Besides external migration, there are also active internal migration processes. The main direction of the migration flows is from the impoverished South to the more prosperous Northern regions, in particular to the capital city, Bishkek. Since independence, the population of Bishkek has almost doubled because of this huge inflow of migrants. Southern regions lack any kind of employment; they are overpopulated and there is a great shortage of land. Very often this internal migration serves as an intermediate stage for earning money to travel to Russia.

The existing studies on migration processes in Kyrgyzstan have looked at the various factors causing migration: economic, social, cultural, and even religious. However, hardly anyone considered the role of environmental degradation in decisions to move. This chapter aims to establish this important missing link and analyse the environmental causes of migration. It has done so through (1) interviews with experts in the field of environment and migration and (2) observations and interviews with migrants as well as the non-migrants still living in the areas affected by sudden environmental disasters and slow environmental degradation.

3 Expert Interviews

One of the goals of this project was to examine the perception of representatives of governmental, non-governmental, and international organisations regarding the association between migration and environmental hazards. Experts were also asked to identify the factors that contribute to – or restrict the likelihood of – environmental migration.

Expert opinion on the role of environmental factors in migration processes was divided; some view migration as a response to high rates of poverty and unemployment and claim that the environmental conditions are of little or no significance. Other experts claim that there is a straightforward connection between environmental hazards (both natural and technological) and migration. They argue that economic factors are intertwined with environmental factors and that it is meaningless and difficult to separate economic from environmental migrants.

Most experts indicated that the majority of residents in risk-prone areas do not migrate. The decision to move or to stay depends upon individual, household, and community characteristics that are discussed later in this chapter. If they do move, usually households move only a short distance away, which is linked to limited resources, family ties, and belief that the effects of hazards are temporary. In addition, there are many cases where migrants eventually return home due to the fact that they could not adapt socially, culturally, or economically to their new environment. Often experts also explain the decision to migrate in terms of the financial situation of the family. The relationship, however, is very complex; some experts argue that migration is an option available only to wealthier households. One argument to support this is that migration involves a number of costs, such as transportation, purchase of new houses in a new place, and purchase of land or livestock. Since poor people cannot afford such expenses, they often decide to stay where they are.

However, a number of experts counter-claim that the less socio-economically advantaged may be more likely to migrate. Families with more assets are believed to be less prone to abandon or sell the house, livestock, and land that they own. On the other hand, poor families have nothing to lose, and therefore, they are more likely to start new lives in new locations.

In addition to material well-being of the households, the association between migration and environmental hazards varies by the setting and location of the household. Thus, most experts underline the fact that among high-risk areas, the rate of out-migration is generally higher in small post-industrial cities and the rural areas of Kyrgyzstan. This can be attributed to limited employment and income-generation possibilities, which lead to poor social and cultural infrastructure of the community.

Migration decisions are also closely related to the strength of community ties. Thus, people are more likely to stay or migrate only temporarily, if most of their relatives reside in the same area. In addition, those who do decide to migrate usually move to places where they have relatives and/or friends. In a country where government support is limited and scarce, social networks are of vital importance. Hence, the presence of relatives and/or friends can ease the process of cultural and social adaptation in a new environment and serve as a source of potential financial and psychological support.

In some cases, only few household members migrate while others stay behind. The decision on who should stay and who should move represents household migratory strategies as related to natural hazards. Many experts have pointed out that those who move tend to be younger or middle-aged and characterised by higher educational levels. This decision is made on the assumption that such family members have a wider variety of employment opportunities in new locations.

Results also indicated that a lack of government aid restricts the residential choices of many inhabitants of risk-prone regions. Most of the experts noted that they were unaware of any centralised programme on environmental migration. Moreover, the rate of out-migration is lower in communities where disaster preparedness and/or relief programmes are carried out.

Many residents in hazardous locations do not migrate because they have no choice. Most of the land in Kyrgyzstan is privatised. The scarcity of land is especially visible in the Southern regions where the majority of natural disasters take place. Those who wish to relocate are either provided with land of 'less quality' or land in other risk-prone areas. Many people refuse to relocate and sign a document stating that they are aware of potential risk and confirm their informed decision to remain in the dangerous places.

Finally, people refuse to move due to emotional attachment to their location, which is more apparent among older inhabitants. They claim that they cannot move since these are the places where they were born and have been raised, and that they would like to be buried with their ancestors in the same places.

Although experts have different perceptions regarding the role of environment in migration decisions, all of them believe that the influence of environmental hazards on different aspects of human life will intensify, and, as a result, the number of environmental migrants will increase. Therefore, many experts suggested the design of a strategic programme that will address the short- and long-term consequences of environmental hazards on human well-being.

Interviews with experts allowed the research team to shape a general perspective on the situation, to formulate some additional questions and issues for our fieldwork, and to identify areas for fieldwork visits. The research then employed methodology developed by the EACH-FOR project, which included observations and interviews with non-migrants and a questionnaire survey of migrants and non-migrants in the identified settlements. The team of four split into two sub-teams: one studied the Naryn province in Northern Kyrgyzstan and another studied the Osh and Jalalabad provinces in Southern Kyrgyzstan. This chapter presents the findings of the fieldwork in two parts described in the following sections: (1) description of settlements and (2) analysis of questionnaires.

4 Researched Settlements

The research team conducted work in seven territories/settlements in Southern and Northern Kyrgyzstan: Min-Kush and Chaek (Naryn province), Mailuu-Suu (Jalalabad province), Kara-Sogot, Nichke-Sai, Kara-Kulja, Kurshab, and Manas Aiyly (Osh province). Min-Kush, Chaek, and Mailuu-Suu were chosen as areas affected by radiation from uranium mines. Other settlements were chosen on the basis of major natural disasters: earthquakes, landslides, and floods. Due to limited space, this chapter describes only two settlements: Mailuu-Suu – a town with uranium wastes, and Kara-Kulja – an area of frequent landslides and floods.

4.1 Mailuu-Suu

Mailuu-Suu is a city in Jalalabad province. For a long time, it was a city with restricted access, not only for foreigners but also for Soviet citizens. Earlier, it was classified as 'secret city' and was called 'Mailbox 200'. It is situated only 150 km from Osh. Mailuu-Suu is marked as the most ecologically disastrous area of Central Asia. The Blacksmith Institute, which does research on environmental issues, placed Mailuu-Suu in the top 30 of the most polluted cities of the world. There are 2 million tons of uranium waste contained in 23 tailing-fields. The full area of the tailing-fields is 432,000 m^2. The combined radioactivity of all the tailing-fields of Mailuu-Suu is 5,000 curie (MES, 2008). Within 22 years (1946–1967), from the work of 2 hydrometallurgical factories, 10,000 tons of uranium oxide and protoxide were extracted. Mailuu-Suu survived a crisis of closing and re-profiling of the uranium pits. Later, an electric light bulb factory was opened in the city. Currently, soil erosion and landslide activity are on the rise for two reasons: the mines in the hills and overgrazing of the hills by local animal stock. Such a situation rings alarms for the Kyrgyz government, for Uzbekistan, and for a broader international community. If tailing-fields are damaged and uranium waste slides into the waters of Mailuu-Suu River, it will lead to a regional ecological catastrophe and poison a large part of the Fergana valley. Currently, the MES is using money from international donors to rebury the waste on the same territory.

The research team spent 1 day in the Sary-Be village on the outskirts of Mailuu-Suu; 17 tailing-fields surround the village. What the team saw looked more like a scene from a war or horror movie (see Fig. 18.1). Some surviving, but empty, four-story apartment buildings were surrounded by the ruins of houses dissembled for construction materials. The majority of the population was elderly. Most of the younger generation has left. Many of those who

Fig. 18.1 Ruins of the Sary-Be village in Mailuu-Suu (source: authors)

remained are alcoholics. One fairly drunk fellow suggested that 'if you have 100 g of vodka every day, it will reduce the effect of radiation'. But for us it looked more as if alcohol offered a means to forget about the terrifying reality of one's body being slowly and continuously destroyed by radiation, which according to local authorities and doctors was a hundred times higher than is safe. Interviews with residents and with a doctor at the local clinic revealed that the effect on health was enormous. One young mother of four described how her mother died from cancer, her father was blind and terminally ill, she had cancer herself, and her four children all had goitres. The doctor mentioned dozens of illnesses common to their village: various types of cancer, low energy, fertility problems for women and potency problems for men, poor mental health for boys, and many other problems. People are simply dying, but still, they stay. Why?

Trying to have this question answered, we were told a local hypothesis – which may or may not be true – that says that the bodies of local residents are used to high radiation to the extent that they cannot survive without it, so those leaving the place either die or return back, since they are unable to adjust to a normal radiological environment. A less-complicated explanation was that they simply did not have means to escape; there was no help from the government and no personal funds to migrate. Finally, in spite of radiation, they had some land for vegetables and fruits and small pastures to feed their animals on the hills. The team left the village with frustration. Most of the team suffered for a week after the trip, either because of the touching scenes or because of only few hours of exposure to high radiation.

4.2 Kara-Kulja

The research team studied several other villages located along Kara-Kulja River, a territory prone to landslides and floods. Major landslides happened in this area 2–3 times since the 1990s. A total of 209 people were relocated from this area. Usually, the government and local authorities monitor the landslides and try to move people to safer places when there are potential dangers. The authorities ban people from using dangerous places and ban animal grazing in those areas. In spite of the danger, some people never move. Others move to safer places for a while but soon return, mainly because the new land they are given is small and lacks water, whereas their original places are abundant in land, water, and grass. Many families move out temporarily from their houses during dangerous seasons like in the spring, or during heavy rains, and then return back.

Sary-Tash (Yellow Stone) is a village of those who relocated permanently because of landslides. All houses here are newly built, but many of them are empty. The village is surrounded by hills on one side and the very steep bank of Kara-Kulja River on another. The government provided long-term loans, which covered most of the construction costs for the new homes. Some construction materials were also provided. The main concern of the villagers was that they were moved from one environmental disaster to two new ones; lack of water and a newly developing landslide. When they were relocated, they were told that the new location was safe to live. However, after they built houses and started to settle, a new landslide emerged on the nearby hills (see Fig. 18.2). This landslide is getting closer and closer each year, but the government does not address the issue, referring to the argument that the displaced people have already received their money and that this landslide will not occur. Lack of water is not considered a sufficient reason to relocate them.

Sary-Tash has no school and basically no services. Residents have no access to clean drinking or irrigation water. It was a dry, dusty, and fairly empty settlement. The villagers complained that locals do not allow them graze their animals on the local pastures, that is why they have to graze their animals in their own pastures near their original settlements, far away from the current place of residence. Migrants in Sary-Tash live in miserable conditions with access to almost nothing. They are cut-off from their relatives and their community, and their basic needs are not met. This new village is already experiencing out-migration as many realise that there is no future in Sary-Tash. Therefore, people are even leaving the newly built houses in search of a more suitable place to live. The case of Sary-Tash village shows how poor planning of new settlements, and a lack of proper land, often leaves those who migrate in worse conditions than before.

The two settlements described above help us form an overall impression of the migration situation and draw some major preliminary conclusions. These two settlements portray two very different images with one common factor: distress. It seems that both those who stay and those who migrate were losing. To test this finding and look for other, deeper connections between environmental problems

Fig. 18.2 New landslide developing near Sary-Tash village in Kara-Kulja region (source: authors)

and migration, we conducted an analysis of migrant and non-migrant surveys. Due to limitations of space, only summaries of the analysis are presented in the following section.

5 Migrant and Non-migrant Questionnaire Survey

5.1 Migrant Survey: The Environmental Migration Chain

The research team studied 30 migrant families living in 11 villages located in 3 provinces of Kyrgyzstan. The analysis produced the following chronological chain of events describing migration caused by environmental problems.

In the mountainous regions of Kyrgyzstan, animal husbandry is the main source of income and survival for the population. Given the harsh economic realities after the break-up of the Soviet Union, people tried to increase stock breeding. In the absence of state-regulated control, this led to overgrazing of pasture lands. Overgrazing led to soil erosion and to increased landslide and flooding activity. Environmental degradation and danger became the main reasons for migration. No other factors (economic, social, or political) carried as much weight. The government supported resettlement, both forcefully and with long-term loans for building houses. They also provided people with new land. However, because of overpopulation in Southern Kyrgyzstan; the new land provided was very small and located in places with little water and no grazing lands nearby. Therefore, migrants often moved from better places to much worse new locations. They used their savings and livestock for migration, only to find that they can barely survive there. In the migration process, they managed to preserve some of their family links but lost almost all broader social networks.

Therefore, at the beginning of the environmental migration chain, we see harsh economic realities and the absence of control over people's activities as the main reason for environmental degradation, forcing people to migrate. At the end of the same chain, there are dislocated migrant families who have lost their natural, social, and cultural environments, as well as most economic opportunities.

5.2 Non-migrant Survey: Living in Ruins

Our research also studied 36 non-migrant families living in 7 villages in 3 provinces in Kyrgyzstan. The results show that those who remained behind are in general economically better off than those who left. They preserve their land and animal stock. However, they are massively affected by environmental disasters/deterioration and by out-migration. The main effects of the environmental problems are on their physical and psychological health (especially in the places of former uranium mines) and on their livelihood (decline of harvest and loss of animal stock). The effect of out-migration results in the lost socio-cultural milieu, ageing population, demoralised youth, and deterioration of physical infrastructure. As many as 80 families leave some villages in 1 year. The pressure for those who stay builds up: most of them want to leave because of environmental problems, but cannot. On one hand, they are afraid of economic difficulties in new locations, and on the other, they do not have the means to move. Hence, they carry on waking up everyday to the horrifying reality of their bodies being destroyed every second of living in these radioactive zones or hoping that their homes will not be squashed by a developing landslide.

6 Conclusion: Scenario with No Winners

The Kyrgyz case shows that migration processes in Kyrgyzstan have a strong environmental flavour and that environmental problems causing migration and displacement of large groups of people bring only negative impacts on the life of local communities. The uniqueness of environmental migration is that very often it is not a choice, but the only solution in a no-win scenario: both those who migrate and those who stay are losers. Their social networks, built over an entire lifetime, are destroyed. Their livelihood strategies, inherited from many generations of forefathers, become meaningless. Their culture, strongly connected to tribal identity and unity, disappears. This study brings into focus the importance of this often-ignored environmental dimension in the study of population dynamics. It also points to the interconnectedness of environmental factors with economic, social, and cultural factors affecting the migration flows of people.

7 Recommendations: Restoring the Balance

As this study has revealed, almost all disasters have anthropogenic roots, whether on the scale of large governmental industries or small-scale individual stockbreeding practices. Current governmental policies in Kyrgyzstan, however, deal mostly with consequences and do not address the real causes. There is a need for both top–down and bottom–up initiatives for preserving the slopes, forests, lakes, and rivers of the Kyrgyz ecosystems. One solution can be found in reviving the traditional indigenous methods of animal husbandry and agriculture, which were developed during several millennia of evolution. Kyrgyz nomads were always very considerate of treating pastures with great care. They were not attached to one place but were following the favourable climate, and thus, were a part of a complex and interconnected ecosystem. Many of these practices and a large part of traditional ecological knowledge were lost over the last century of collective farming and sedentarisation.

It is also important to revive the indigenous breeds of sheep and sorts of fruits and vegetables grown in Kyrgyzstan. These are much better adjusted to local environmental conditions. For example, the indigenous Alay sheep consumes much less grass and exerts significantly smaller pressure on pastures while feeding, as compared to many new heavy hybrid breeds brought and cultivated in Kyrgyzstan during the Soviet times. Another example is the indigenous *Kasek* apricot which requires watering only 2–3 times a year, in contrast to many imported apricot trees requiring 6–7 times more water. This is extremely practical in the Southern Batken province strongly affected by droughts. Currently, there are several initiatives taking place in Kyrgyzstan, aimed at preserving the agro/biodiversity of the region by using the indigenous methods. Reviving the traditional wisdom and practices could help many Central Asian communities restore the lost ecological balance, which will be the first step towards restoring the damaged social and cultural milieu of the Kyrgyz settlements and communities suffering from forced migration.

Acknowledgements This research was carried out within the framework of the EACH-FOR Project with the financial support of the European Commission's Sixth Framework Programme (contract number 044468).

References

Ministry of Emergency Situations. (2008): Natural and technogenic emergency situations, MES.

Stepanenko, I. (1997): Природно-климатические условия Кыргызстана, ЦАИК.

Radjabov, M. (2008): Проблемы региональной безопасности в Центральной Азии на пороге XXI века, Газета «Навигатор» (Казахстан).

Zvyagelskaya, I.D. (2001): К вопросу об угрозах в Центральной Азии, Материалы научно-практического семинара «10 лет СНГ: некоторые итоги», СЕРИЯ: Научные доклады № 5. Июль 2001.

Gruzdov, Y. (2003): Где хвост уранового дракона? Кыргызстан, №11 за март 2003.

Mamajakyp uulu, C., E.B. Kamcgybeov, C.M. Magazova, D. Nazarbekov, A.K. Nartaeva, and S.M. Mamatov (2005): Изменение клеток периферической крови у жителей, проживающих в районе захоронения радиоактивных отходов, Научно-исследовательский ледовательский отдел ДМСРЦ Министерства здравоохранения Кыргызской Республики, Кыргызско-Российский Славянский Университет (г. Бишкек).

Tishin, A.I. (2007): Процессы социальной диссипациив Кыргызской Республике, Кыргызский Национальный университет, г. Бишкек.

Chapter 19
Linking the Earth's Future to Migration: Scenarios of Environmental Change and Possible Impacts on Forced Migration

Johannes Frühmann and Jill Jäger

1 Introduction

This chapter discusses how to develop scenarios of environmental change to estimate future migration flows that are forced by environmental degradation. The methodology as well as the examples presented here was developed in the EU-funded project EACH-FOR (www.each-for.eu). In the next section, the chapter discusses environmental degradation and gives an overview of the 'hot spots' in the case studies. This section is followed by an introduction to scenarios in general and the scenario narratives of the fourth Global Environmental Outlook (GEO 4; UNEP and United Nations Environment Programme, 2007) and the Intergovernmental Panel on Climate Change (IPCC) which were used in the EACH-FOR project. The fourth section gives a detailed description of the EACH-FOR approach – including two case study examples from Egypt and Mozambique. The chapter finishes with conclusions that sum up the first results of the work with scenarios in the case studies and reflect on the approach taken in the EACH-FOR project.

2 Hot Spots of Environmental Degradation

2.1 Environmental Degradation

Environmental degradation is the deterioration of the environment through depletion of resources, such as air, water, and soil; the destruction of ecosystems; and the extinction of species. *The United Nations Environment Programme (UNEP)'s Global Environmental Outlook* (UNEP and United Nations Environment Programme, 2007) concludes that there is evidence of unprecedented environmental change at global and regional levels. These changes include the warming of the Earth's surface with associated changes in water availability,

J. Frühmann (✉)
Sustainable Europe Research Institute, Vienna, Austria
e-mail: johannes.fruehmann@seri.at

T. Afifi, J. Jäger (eds.), *Environment, Forced Migration and Social Vulnerability*,
DOI 10.1007/978-3-642-12416-7_19,

melting of snow and ice, rising global-average sea level, and loss of biodiversity. Unsustainable land use and climate change are driving land degradation including soil erosion, nutrient depletion, water scarcity, salinity, desertification, and the disruption of biological cycles. The per-capita availability of freshwater is declining globally and contaminated water remains the greatest single environmental cause of human sickness and death. Aquatic ecosystems continue to be heavily exploited with a large-scale decline in fish catches. While air pollution has been controlled in some areas, it continues to increase in many cities and indoor air pollution due to the improper burning of solid biomass fuels continues to impose an enormous health burden.

2.2 Case Studies

For the EACH-FOR project a series of case studies were carried out in order to explore the links between environmental degradation and migration. As a first step, the main types of environmental degradation or environmental stress in each of the case study areas were identified.

A survey of the case studies suggests four general patterns of environmental change to be explored in terms of potential impacts on migration. First, many of the case study areas discuss natural disasters, in particular floods and droughts. The Fourth Assessment Report of the IPCC (IPCC and Intergovernmental Panel on Climate Change, 2008) notes that confidence has increased that some extreme weather events will become more frequent, more intense, and more widespread in the twenty-first century. This has implications for migration. Secondly, almost all of the case studies include land degradation and/or desertification. These are slow processes resulting from a number of causes, including poor soil management, overfertilisation, overwithdrawal of groundwater, and removal of protective vegetation cover, as well as climatic change. As *GEO 4* (UNEP and United Nations Environment Programme, 2007) notes, poor people suffer disproportionately from the effects of land degradation, especially in the drylands which support some 2 billion people. Third, two of the case studies focus on sea-level rise. The IPCC notes that coastal zones are likely to be exposed to increasing risks, including coastal erosion due to climate change and sea-level rise. Furthermore, the IPCC states that 'Many millions more people are projected to be flooded every year due to sea-level rise by the 2080s.' (IPCC and Intergovernmental Panel on Climate Change, 2008: 7). Fourth, three of the cases focus on forced migration as a result of dam-building projects. The driving factors behind such projects often have an environmental origin when the dams are built in order to provide water for irrigation.

In addition to exploring possible current links between environmental degradation, natural hazards, and development projects on one hand and forced migration on the other, it is of interest to discuss how these links and their consequences could evolve in the future. This can be done using scenarios.

3 Scenarios

3.1 *What Is a Scenario?*

A scenario is not a prediction of what the future *will* be but rather a description of how the future *might* unfold. Scenarios explore the possible, not just the probable, and challenge their users to think beyond conventional wisdom.

Scenarios are carefully created stories about the future. A scenario includes an interpretation of the present, a vision of the future, and an internally consistent account of the path from the present to the future. Scenarios support informed action by providing insights into the scope of the possible. They also can illustrate the role of human activities in shaping the future and the links among issues. In the process of helping to clarify possible future developments and their effects, scenarios often are a source of inspiration for creative ideas. Jäger et al. (2008) describe the development and use of scenarios in integrated environmental assessments.

3.2 *The GEO 4 Scenarios*

In 2007 the United Nations Environmental Programme (UNEP) published their fourth Global Environmental Outlook (GEO 4; UNEP and United Nations Environment Programme, 2007). The report presents four different storylines in order to explore the earth's future with a scenario-based approach: Markets First, Security First, Policy First, and Sustainability First. The scenarios are based on the consultation of many stakeholders and experts on both global and regional levels. First, the storylines were developed at the global level; in a second step, the global narratives have been applied to the different contexts on the regional level. In this section, the global storylines are briefly summarised.

Under *Markets First*, the private sector widens its influence into previous governmental areas, i.e. privatisation of education, security, research, health, and other social services, together with a continued movement towards free trade and the commoditisation of nature. International trade accelerates, although no global free-trade zone is achieved. Existing regional agreements are strengthened and new South–South cooperations emerge. Formal environmental protection is limited by efforts to increase economic investment and expand trade. The Kyoto Protocol sees no significant international follow-up after 2012. Ecosystem services are turned into commodities. The economic exchange of goods like water, genetic material, knowledge, and culture increases dramatically, while terrestrial and marine biodiversity decrease.

Agriculture is intensified in all regions, which increases the potential of soil degradation. This could lead to problems especially in Latin America, the Caribbean, and Africa, where the intensification is not accompanied by a net reduction in cropland. Water-use efficiency increases in most regions (due to privatisation and better technologies), but the number of people living with severe water stress grows significantly because of growing populations and climate change (UNEP and United Nations Environment Programme, 2007).

Security First can be also described as 'Me First' and brings a fairly narrow notion of security that implies increased limits on how people live, both physically and physiologically. Restrictions on migration reduce the movement of people and trade barriers that of goods. Governments are strong in decision-making, but multinational corporations and private interests increase their influence. The authority of international institutions declines and public participation gets marginalised. Total energy use increases significantly, while energy efficiency slowly improves. A dramatic resurgence in the use of coal results in strongly rising levels of atmospheric CO_2. The combination of climate change, growing populations, and greater economic activity strains freshwater resources (both quantity and quality) and brings a dramatic increase of people facing severe water stress; conflicts on shared resources result. Both terrestrial and marine biodiversity are under great pressure (UNEP and United Nations Environment Programme, 2007).

The *Policy First* scenario is a highly centralised approach with a move to a more holistic government balancing economic growth with social and environmental issues. National governments and international organisations (e.g. United Nations) lead in those efforts. Subsidies that encourage the overexploitation of resources are reduced or eliminated. Public investments in science and technology grow, and the number of protected areas increases with more or less effective efforts in preventing land-use change in these areas. International agreements increase energy efficiency and induce a move to more low carbon and renewable sources (i.e. biofuels). However, total energy consumption continues to increase. The higher demand for biofuels results in a significant increase in land devoted to pasture and a decrease of forest land. Growing populations and economic activities still put pressure on resources, particularly in developing regions. The number of people living under severe water stress continues to rise, but institutional efforts to better share resources help limit the impacts. The increased demand, however, places a strain on the quality of water resources. The loss of biodiversity is dramatic, mainly due to climate change and agricultural practices (UNEP and United Nations Environment Programme, 2007).

In *Sustainability First*, actors from all levels (local/national/regional/international) and from all sectors (government/civil/private) address environmental and social concerns. The rules of international trade are reformed and public resources are shifted from military to social and environmental issues. Climate change remains a persistent problem. Even though the growth in the level of CO_2 is limited, it is not possible to avoid potentially significant warming and sea-level rise. In the energy sector, total energy use increases, but the mix of fuels changes significantly with wind, solar, and modern biofuels becoming an important fraction beside natural gas as the dominant source. The expansion of agricultural land comes at the expense of forest land, but the loss of the latter slows significantly over time. The growth of water stress is reduced, but some regions still face increasing problems, both with quantity and quality. Efforts to reduce biodiversity loss are high, but the challenges are strong and there is significant species loss in some areas (UNEP and United Nations Environment Programme, 2007).

3.3 The IPCC Scenarios

IPCC used narrative storylines for the development of quantitative scenarios, i.e. climate projections. The narratives describe the relationships between the drivers of greenhouse gas (GHG) and aerosol emissions and their development until 2100. The storylines were written for large world regions and a global scale. They differ in demographic, social, economic, technological, and environmental lines of development that result in different levels of GHG emissions. Furthermore, the storylines assume that no specific climate policies are implemented, and thus, they form a baseline against which narratives with specific mitigation and adaptation measures (as the GEO 4 narratives) can be compared.

Basically, the IPCC storylines differ along two axes: Global versus regional and economic versus environmental emphasis. The A1/B1 narratives are more

Table 19.1 Summary characteristics of the scenario storylines used in the IPCC scenarios (according to Carter et al., 2007)

	A1b	A2	B1	B2
World	Market-oriented	Differentiated	Convergent	Local solutions
Economy	Fastest per capita growth	Regionally oriented; lowest per capita growth	Service and information based; lower growth than A1	Intermediate growth
Population	2050 peak; then decline	Continuously increasing	2050 peak; then decline	Continuously increasing at lower rate than A2
Governance	Strong regional interactions; income convergence	Self-reliance with preservation of local identities	Global solutions to economic, social, and environmental sustainability	Local and regional solutions to environmental protection and social equity
Technology	Balanced across all (energy) sources	Slowest and most fragmented development	Clean and resource-efficient	More rapid than A2; less rapid and more diverse than A1/B1

global, whereas the A2/B2 narratives are more regionally oriented; The A1/A2 have more economic emphasis, while the B1/B2 have more environmental emphasis. More details to differentiate the narratives are given in Table 19.1.

3.4 Linking Different Scenarios

The distinction between global/regional and economic/environmental makes it quite easy to relate the GEO 4 narratives to the IPCC narratives (Fig. 19.1). The combination of different scenarios offers more insights into possible future developments. For the work of the EACH-FOR project, the IPCC climate projections offer detailed information on changes in temperature and rainfall patterns. This is an important addition to the GEO 4 scenarios, especially in case studies focused on desertification and/or water shortage.

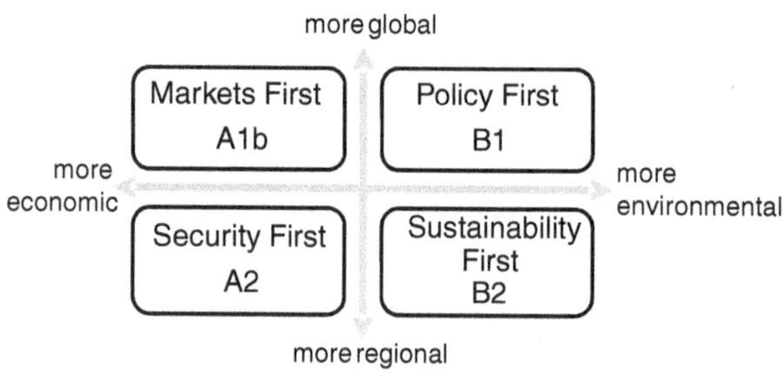

Fig. 19.1 Linking narratives from GEO 4 and IPCC (figure by the authors, based on the following sources: Carter et al., 2007; MA and Millennium Ecosystem Assessment, 2003; UNEP and United Nations Environment Programme, 2007)

4 The EACH-FOR Approach

4.1 Methodology

Environmental degradation and migration patterns differ from one region to the other. Therefore, the methodological approach of the EACH-FOR project is designed to be as flexible as possible and geographically based to be able to tackle the (regionally or locally) different research questions. In total, there are 23 case studies.

The EACH-FOR project was essentially carried out in four steps: In the first step, general overview studies were conducted to describe the existing situation regarding environmental problems and migration for each region as a background to the selection of the case studies. In the second step the field work in the case studies was carried out. The third step comprises the work on forced migration scenarios (described in detail below), before in the last step key findings for policymakers can be formulated.

In order to generate scenarios for environmentally forced migration, we considered a three-step approach that aims for a participatory scenario development:

(1) Develop 'scenario input papers' on environmental degradation for six selected case studies.
(2) Consortium partners that carried out the case studies add specific information on migration and develop forced migration scenarios.
(3) Discuss these results with local experts in each case study to refine the scenarios.

The 'scenario input papers' were developed with a consistent outline to generate local scenarios that are comparable. They are based on the scenarios of GEO 4 and IPCC and other relevant studies with a regional or local focus and take a multilevel approach. The papers start with the global perspective of the GEO 4 narratives, where two contrasting storylines are selected; one of the pair Markets First/Security First and one from Policy First/Sustainability First. For the regional level, the implications of the two storylines with an emphasis on the significant hot spots of the case study are developed. The national level is then covered by a set of indicators (see Table 19.2) derived from International Futures (IFs and International Futures, 2008). The indicators show divergent developments under different scenarios and are interpreted according to the focus of each case study. The IPCC climate projections are added on the regional level to show changes in temperature and rainfall patterns. Finally, other studies that contain relevant scenarios for the case study area are included in the input papers (e.g. the African Environment Outlook 2 (UNEP and United Nations Environment Programme, 2006), reports from the Environment and Security Initiative (ENVSEC and The Environment and Security Initiative, 2008), and others).

Table 19.2 Set of indicators used from international futures (IFs and International Futures, 2008)

Indicator	Unit
Yield	Tons per hectares
GDP per capita at PPP	Thousand $
Land (forest cover)	Million hectare
Water usage, annual	Cubic km
Aid (foreign), net	Billion $
Population	Million people
Life expectancy	Years
Malnourished children	Percent
Physical quality of life index	Index (max. 100)

Each 'scenario input paper' is then complemented by the consortium partner that led the case study. Often this is done in cooperation with local scientists in the case study area that supported the field work. This process aims to relate the specific information on migration to the scenarios and results in integrated environmental-migration scenarios.

Those scenarios can then be discussed with local experts – preferably in a workshop setting – for further refinement of the scenarios. Including the view of local stakeholders is in fact very important for applying the storylines to the local/regional context. However, this step is expensive and time-consuming and was essentially not possible within the limited budget, time, and structure of the EACH-FOR project.

4.2 Example 1: Mozambique

4.2.1 Brief Introduction to the Case Study Area

Mozambique is located in southeastern Africa, with a 2,500-km long coastline along the Indian Ocean, between South Africa and Tanzania.[1] About 20 million people live in the country that covers an area of 800,000 km^2. The climate varies from tropical and subtropical in the north to semiarid and dry in the south. Temperatures are high, and the average annual precipitation is 1,032 mm, with a rainy season from October to April. Variations in rainfall range from 2,000 mm in the north and central regions to less than 400 mm in the southern inland (FAO and Food and Agriculture Organization of the United Nations, 2007; Stal, 2009; UNU-EHS, 2008).

The hot spots of environmental degradation we focused on were floods and droughts. For example, in 2000, the Limpopo River swelled up to 80 km wide and caused a huge flood for more than 3 months that affected 4.5 million people and destroyed their livelihoods, seed reserves, and crop fields. After a smaller flood event in 2003, the latest flood of the Zambezi River in 2007 affected 500,000 people, when 6 major cyclones hit the country.

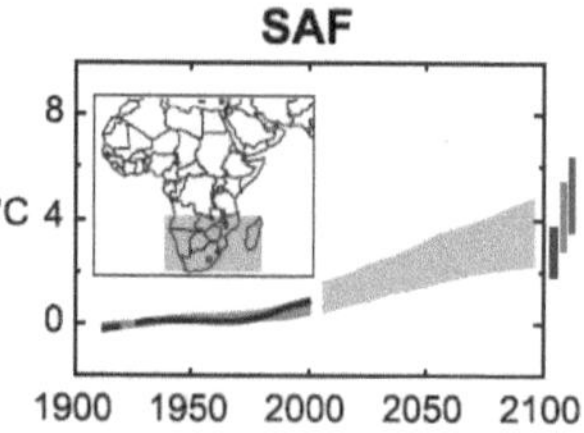

Fig. 19.2 Temperature anomalies with respect to 1901–1950 for South Africa for 1906–2005 (*black line*) and as simulated (*grey* envelope); and as projected for 2001–2100 by MMD models for the A1B scenario (*grey* envelope). The bars at the end of the *grey* envelope represent the range of projected changes for 2091–2100 for the B1 scenario (*left*), the A1B scenario (*middle*), and the A2 scenario (*right*) (Christensen et al., 2007)

[1] For further details see UNU-EHS (2008) and Stal (2009), both sources available on www.each-for.eu.

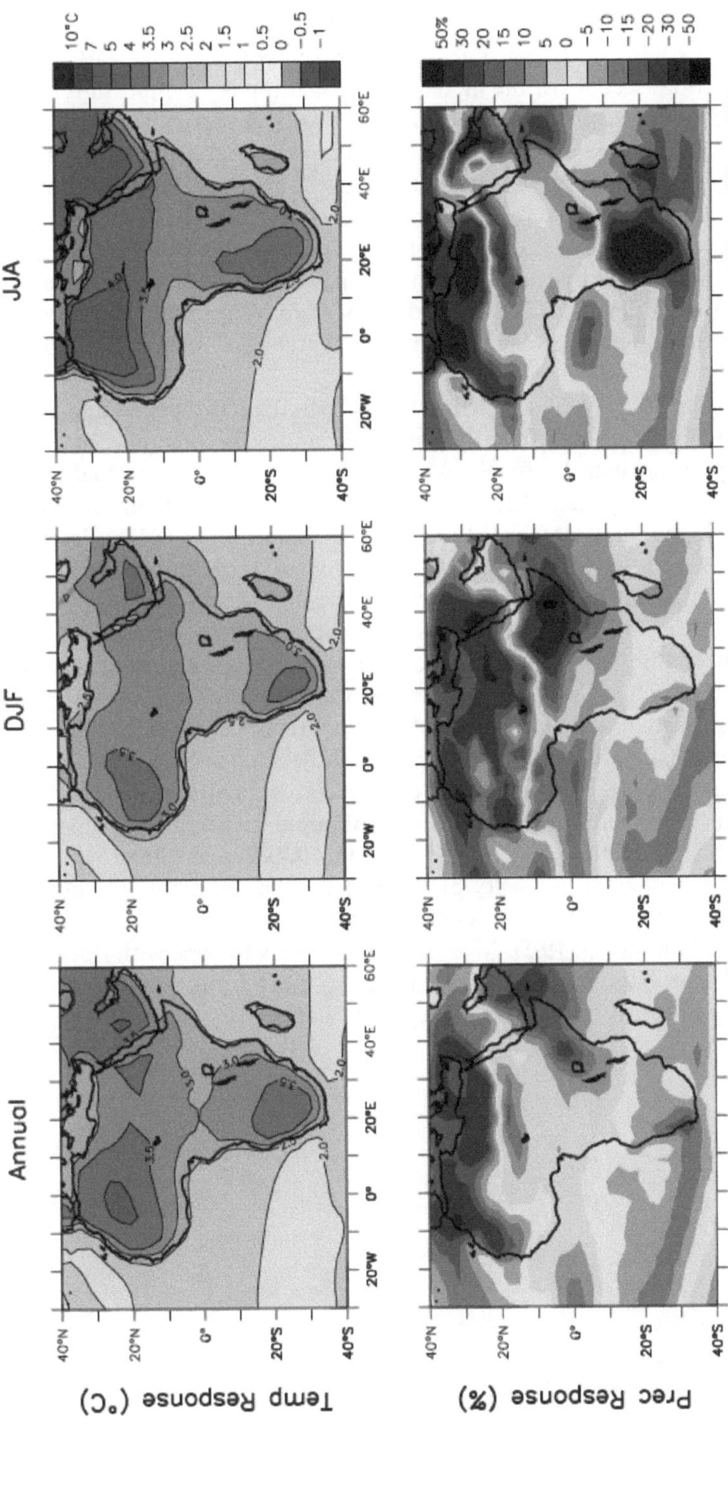

Fig. 19.3 Temperature and precipitation changes over Africa from simulations of the A1b scenario. *Top row*: Annual mean, December/January/February and June/July/August temperature change between 1980–1999 and 2080–2099, averaged over 21 models. *Bottom row*: Same as *top*, but for fractional change in precipitation (Christensen et al., 2007)

4.2.2 Scenarios

For this case study, we decided to use the *Markets First* and *Policy First* scenarios. In both scenarios, the increasing population is an important driver. Under *Markets First*, an intensified and profit-driven agriculture is characterised by unsustainable practices and leads to severe land degradation. Privatisation and amalgamation of sectors improves human development, but globalisation and limited efforts for environmental issues have negative impacts. In *Policy First*, regional integration and policies with economic and environmental stewardship are guided by the regulatory frameworks of the New Partnership for Africa's Development (NEPAD) and the African Ministerial Conference on the Environment and help reaching environmental and human development goals.

According to the IPCC scenarios, the potential of (summer) floods and (winter) droughts rises significantly, especially under the A1b narrative. The median temperature increase for the A1b scenario is about 1.5 times higher than the global mean response. As shown in Fig. 19.2, the temperature rise is significantly higher under Market First (= A1b) conditions by 2100. Precipitation decreases slightly – at least in the A1b scenario as displayed in Fig. 19.3. The main changes in precipitation occur in the dry season which has minor effects on the overall water resources, but could lead to more droughts – especially in combination with the increase of the frequency of dry austral winters of about 20%. At the same time, the frequency of extremely wet summers is seen to double in southern Africa (Christensen et al., 2007).

On the national level, the IFs indicators show similar trends under both scenarios, but there are some important differences. Agriculture yields rise more under *Markets First*, but the slower increase of population almost compensates the lower food productivity in *Policy First* (Fig. 19.4). Under the projection of increasing frequency and intensity of extreme weather events in combination with the high vulnerability of the local population, another indicator becomes crucial: Foreign Aid (Fig. 19.5). Mozambique's ability to cope with floods and droughts will probably depend on financial support from outside, and this support is significantly higher in a *Policy First* world.

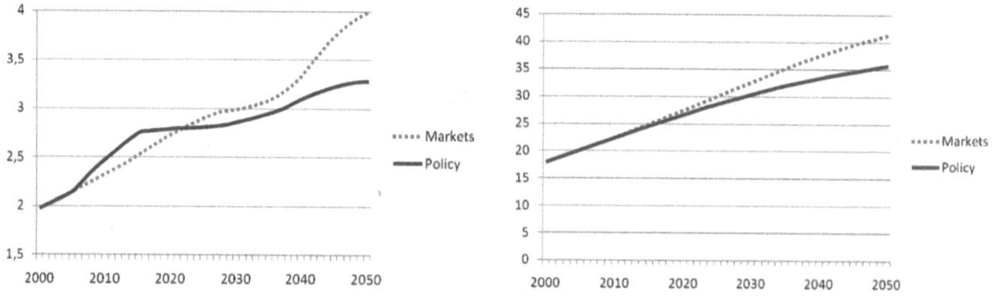

Fig. 19.4 Yields in tons per hectares (*left*) and population in million people (*right*) in Mozambique. (Figures derived from data from IFs and International Futures, 2008.)

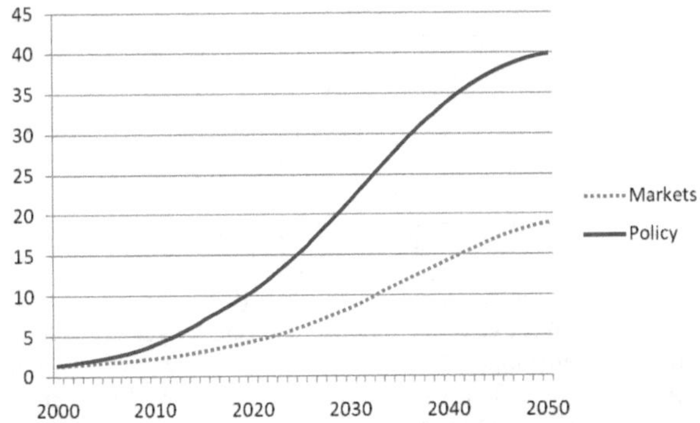

Fig. 19.5 Net foreign aid in billion US dollars in Mozambique (Figures derived from data from IFs and International Futures, 2008.)

4.3 Example 2: Egypt

4.3.1 Brief Introduction to the Case Study

Egypt lies in the very northeast of Africa, with an area of 1 million km^2 and about 80 million inhabitants.[2] Hot dry summers and moderate winters with very low annual rainfall (ranging from 200 mm in north to almost zero in the south) characterise the desert type of climate. As the main source of freshwater, the Nile River is 'the lifeblood of Egypt' (EIA and Energy Information Administration, 2003). The river also contributes about 20% of the country's energy and is important for transportation. Egypt's development was and still is strongly related to the Nile River (ERASMUS, 2008). With regard to development and migration, Egypt is of special interest to the European Union (EU) because of its close geographical position.

Water shortage and desertification are big issues today, but in the next decades sea-level rise could become the biggest challenge. The highly populated Nile Delta is especially vulnerable, because it is the most important agricultural land of the country but lies widely below the mean sea level.

4.3.2 Scenarios

In the Egypt case, we compared the *Security First* and *Policy First* scenarios. *Security First* is – especially in Africa – characterised by poor economic policies that lead to overexploitation of water, land, and mineral resources. In contrast, responsibility for social equity and the environment is taken in the *Policy First* scenario, and regional integration is strengthened. Population increases significantly more in the *Security First* scenario, while the Gross Domestic Product

[2] For further details see ERASMUS (2008) and Afifi (2009), both sources available on www.each-for.eu. See also Chapter 15 in this book.

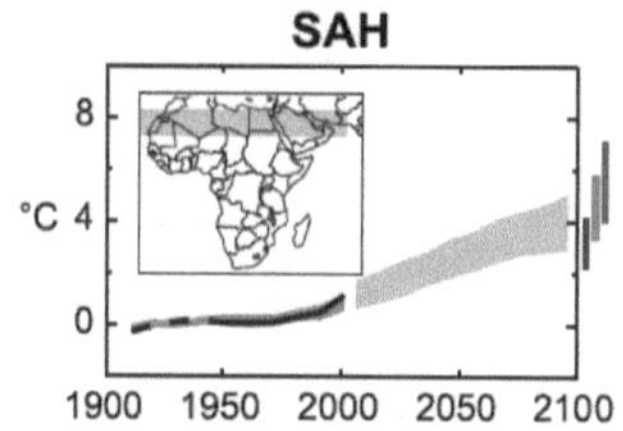

Fig. 19.6 Temperature anomalies with respect to 1901–1950 for North Africa for 1906–2005 (*black line*) and as simulated (*grey* envelope); and as projected for 2001–2100 by MMD models for the A1B scenario (*grey* envelope). The bars at the end of the envelope represent the range of projected changes for 2091–2100 for the B1 scenario (*left*), the A1B scenario (*middle*), and the A2 scenario (*right*) (Christensen et al., 2007)

(GDP) per capita increases at a lower rate than in *Policy First*. This leads to only a small decline in the population with income less than 1 US$ per day and a slightly increasing level of childhood nutrition by 2050.

Figure 19.6 shows big differences between the A2 (Security First) and B1 (Policy First) scenarios in the Saharan region. Under A2 conditions, the mean temperature rises at least by 4°C, which is about the maximum temperature anomaly in the B1 (Policy First) world in 2100. The maximum rise in A2 could even reach 7°C. Figure 19.3 shows that the main temperature rise takes place in the already hot summer months. Furthermore, Egypt faces a dramatic drying of about 20% in 2100, especially in its northern parts – which comes along with a drying pattern in the whole Mediterranean region. But a differentiated look at the precipitation maps shows that the Nile basin is not affected by the drying pattern at all, but it remains unsure, whether this increases the water resources of Egypt. Even though the maps in Fig. 19.3 are based on the A1b scenario, we can estimate that the effects shown will be somewhere between A2 (more extreme) and B1 (less dramatic).

On the national level, both scenarios show very similar trends, both paths bring a better overall quality of life for Egypt. But, for all indicators, *Policy First* conditions bring more improvements – for some indicators, the differences are large. For example, foreign aid increases much more in a *Policy First* world – as also found for Mozambique.

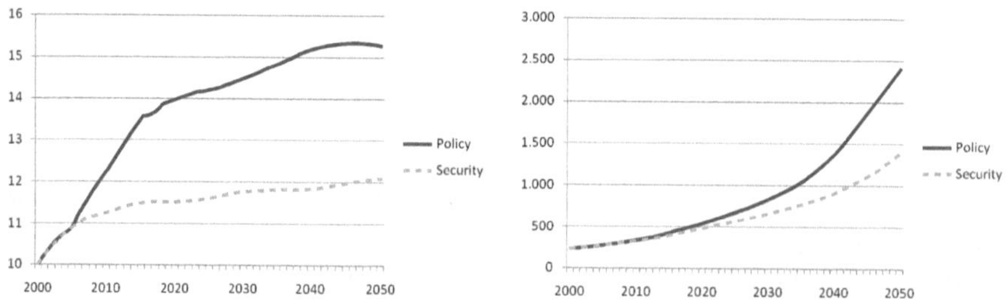

Fig. 19.7 Yield in tons per hectares (*left*) and GDP per capita at PPP in thousand US$ in Egypt (Figures derived from data from IFs and International Futures, 2008.)

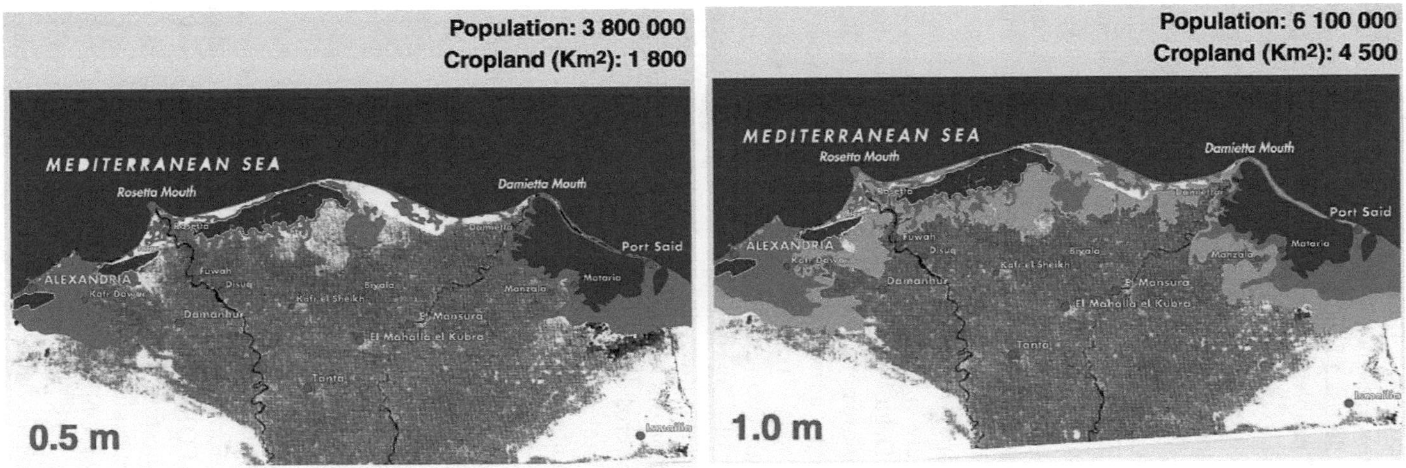

Fig. 19.8 Potential impact of sea level rise in the Nile Delta on population and cropland with a rise of 0.5 m (*left*) and 1.0 m (*right*) (UNEP and United Nations Environment Programme, 2002)

Figure 19.7 shows two more examples: The fast-rising food productivity under Policy First can cope with the increasing population and probably even increase food exports towards the EU and support the growth of GDP. However, it also increases the risk of human-induced land degradation and could lead to environmental problems. From this perspective, also the New Valley Project and other plans to pipe water to desert areas for creating agricultural land have to be studied critically. According to IPCC reports (Boko et al., 2007), Egypt is one of the African countries most vulnerable to climate change because of the impact on water availability. Already in 2000, water use exceeded the available resources of freshwater. So the IPCC report estimates that temperature rises will be likely to reduce the productivity of major crops and increase their water requirements, thereby directly decreasing crop water-use efficiency. At the same time, water stress in all sectors increases due to the projected population increase. There is a high degree of uncertainty about the flow of the Nile, and the ongoing expansion of irrigated areas will reduce the capacity of Egypt to cope with future fluctuations.

However, the possibly biggest future challenge for Egypt is sea-level rise, especially for the Nile Delta. According to the results of the field work (Afifi, 2009), even local experts are not always conscious of this risk. But IPCC reports (Bindoff et al. 2007) project a rise of the global sea level between 0.22 and 0.44 m (in comparison to the sea levels of 1990) by the end of this century. The biggest increase is projected under the A1B scenario. The Nile Delta is especially sensitive, because most of a 50-km wide land strip along the coast is less than 2 m above sea level and is protected from flooding by a 1–10-km wide coastal sand belt only, shaped by discharge of the Rosetta and Damietta branches of the Nile. In recent years, erosion of the protective sand belt has become a serious problem and has accelerated since the construction of the Aswan Dam as the dam holds back sediments from the river water that used to add material to the sand belt in the river delta (UNEP and United Nations Environment Programme, 2002). Figure 19.8 shows possible impacts of 0.5 m and 1.0 m sea-level rise scenarios.

5 Conclusions

For the case studies in the EACH-FOR project, floods and droughts, land degradation, and sea-level rise are the main environmental forcing factors for migration. In addition, migration results from large-scale development projects, and two EACH-FOR case studies examine the effects of building large dams on rivers. The question that we have addressed here is, How could the environmental, economic, and social patterns look in the future? We have used scenarios developed for UNEP's Global Environmental Outlook as well as those developed by the IPCC and, where relevant, these have been supplemented by other scenarios. This chapter provides examples of scenarios for two of the EACH-FOR case studies: Mozambique and Egypt. For Mozambique, we compare the *Markets First* world with the *Policy First* world. In *Markets First* Mozambique experiences significant warming and an increase of winter droughts. The temperature increase is not as high in the *Policy First* scenario. An important difference between the two scenarios is the availability of foreign aid, which is significantly higher in the *Policy First* scenario. The larger climatic changes and lower availability of foreign aid suggest that there would be more migration in the case of the *Markets First* scenario. For Egypt, we compare the *Policy First* and *Security First* scenarios. Again, there are important differences between the scenarios. For example, the agricultural yield is around 25% higher in the *Policy First* scenario. However, the biggest future challenge for Egypt could be sea-level rise, especially for the Nile Delta.

The examples shown here demonstrate how plausible stories about the future can be developed for case study regions. The scenario descriptions do not, however, include estimates of migration rates. The latter need to be derived with the help of local experts. The next step would therefore be to use expert workshops and discuss the scenario narratives, asking the question: What would happen to migration in Scenario A versus Scenario B? The expert input would be used to refine the narratives and provide some quantification of the migration levels. Furthermore, the scenarios can be used in the discussion of policy options. On one hand, these scenarios illustrate in broad terms the differences between taking a 'markets-based' approach versus an 'environmental protection-based' approach, and on the other, they can be used to guide discussions on 'What kind of a future do we want?' and 'What do we need, in order to reach that future?'

Acknowledgements This research was carried out within the framework of the EACH-FOR Project with the financial support of the European Commission's Sixth Framework Programme (contract number 044468).

References

Afifi, T. (2009): Egypt. Case Study Report. EACH-FOR, Environmental change and forced migration scenarios, Deliverable 2.5.2.1, www.each-for.eu

Bindoff, N.L., J. Willebrand, V. Artale, A. Cazenave, J. Gregory, S. Gulev, K. Hanawa, C. Le Quéré, S. Levitus, Y. Nojiri, C.K. Shum, L.D. Talley, and A. Unnikrishnan. (2007): Observations: Oceanic climate change and sea level. In: Solomon, S., Qin, D., Manning, M., Chen, Z., Marquis, M., Averyt, K.B., Tignor, M., and Miller, H.L. (eds.), *Climate Change 2007: The Physical Science Basis. Contribution of Working Group I to the Fourth Assessment Report of the Intergovernmental Panel on Climate Change*. Cambridge University Press, Cambridge, UK and New York, NY.

Boko, M., I. Niang, A. Nyong, C. Vogel, A. Githeko, M. Medany, B. Osman-Elasha, R. Tabo, and P. Yanda. (2007): Africa. In: Parry, M.L., Canziani, O.F., Palutikof, J.P., van der Linden, P.J. and Hanson, C.E., (eds.), *Climate Change 2007: Impacts, Adaptation and Vulnerability. Contribution of Working Group II to the Fourth Assessment Report of the Intergovernmental Panel on Climate Change*. Cambridge University Press, Cambridge, UK. pp. 433–467.

Carter, T.R., R.N. Jones, X. Lu, S. Bhadwal, C. Conde, L.O. Mearns, B.C. O'Neill, M.D.A. Rounsevell, and M.B. Zurek. (2007): New assessment methods and the characterisation of future conditions. In: Parry, M.L., Canziani, O.F., Palutikof, J.P., van der Linden, P.J., and Hanson, C.E., (eds.), *Climate Change 2007: Impacts, Adaptation and Vulnerability. Contribution of Working Group II to the Fourth Assessment Report of the Intergovernmental Panel on Climate Change*. Cambridge University Press, Cambridge, UK. pp. 133–171.

Christensen, J.H., B. Hewitson, A. Busuioc, A. Chen, X. Gao, I. Held, R. Jones, R.K. Kolli, W.-T. Kwon, R. Laprise, V. Magaña Rueda, L. Mearns, C.G. Menéndez, J. Räisänen, A. Rinke, A. Sarr, and P. Whetton. (2007): Regional climate projections. In: Solomon, S., Qin, D., Manning, M., Chen, Z., Marquis, M., Averyt, K.B., Tignor, M., and Miller, H.L. (eds.), *Climate Change 2007: The Physical Science Basis. Contribution of Working Group I to the Fourth Assessment Report of the Intergovernmental Panel on Climate Change*. Cambridge University Press, Cambridge, UK and New York, NY.

EIA, Energy Information Administration. (2003): Egypt: Environmental issues. Available from http://www.eia.doe.gov/emeu/cabs/egypenv.html

ENVSEC, The Environment and Security Initiative. (2008): Environment and security – Transforming risks into cooperation. Available from http://www.envsec.org

ERASMUS. (2008): General overview study Middle East and Northern Africa. EACH-FOR, Environmental change and forced migration scenarios, Deliverable 2.5.1, www.each-for.eu

FAO, Food and Agriculture Organization of the United Nations. (2007): AQUASTAT. Countries and regions: Mozambique. Available from http://www.fao.org/nr/water/aquastat/countries/mozambique/index.stm

IFs, International Futures. (2008): International futures. Xxx. www.internationalfutures.org

IPCC, Intergovernmental Panel on Climate Change. (2008): Fourth assessment report, Summary for policy makers, Working Group 2. www.ipcc.ch

Jäger, J., D. Rothman, C. Anastasi, S. Kartha, and P. van Notten. (2008): *Scenario Development and Analysis*. GEO Resource Book, IISD, Canada and UNEP, Nairobi.

MA, Millennium Ecosystem Assessment. (2003): *Ecosystems and Human Well-Being: A Framework for Assessment*. Island Press, Washington, DC.

Stal, M. (2009): Mozambique. Case study report. EACH-FOR, Environmental change and forced migration scenarios, Deliverable 2.4.2.2, www.each-for.eu

UNEP, United Nations Environment Programme. (2002): Vital Climate Graphics Africa. The Nile Delta study. Available from http://www.grida.no/climate/vitalafrica/english/16.htm

UNEP, United Nations Environment Programme. (2006): African Environment Outlook 2. Our environment, our wealth. Available from http://www.unep.org/DEWA/Africa/AEO2_Launch/

UNEP, United Nations Environment Programme. (2007): Global Environmental Outlook. GEO4. Environment for development. progress press, Malta. Available from http://www.unep.org/geo/geo4/media/

UNU-EHS. (2008): General Overview Study Sub-Saharan Africa and Ghana. EACH-FOR, Environmental change and forced migration scenarios, Deliverable 2.4.1, www.each-for.eu

The Bonn Points

'Bonn Points' is a play on the French phrase for 'good point' (*les bons points*), and the name of the city hosting the Environment, Forced Migration, and Social Vulnerability (EFMSV) International Conference, 9–11 October, Bonn, Germany. The idea was initiated by UNU-EHS to capture the answers of the EFMSV participants to the critical questions that required examination during the conference.

In order to streamline the debate on environmental migration, nine questions were formulated and structured in three thematic groups. The following answers are based on feedback from the participants of the conference. Naturally, the almost 300 conference participants from various scientific backgrounds were unlikely to agree on succinct consensus answers to the complex questions. Thus, the answers may sometimes be formulated as provocative statements triggering further thoughts and debate rather than concluding the debates. We hope that the 'Bonn Points' effectively distill the spirit of the conference, highlight its findings, and indicate 'terra incognita' for further research on *environmentally induced migration*.

Definitions and Conceptualisation

1. What can we call people migrating due to environmental reasons? What if any typology could capture the spectrum of environmentally related migration, including the severity of the environmental stressor and the urgency of the resulting migration decision?

The conference did not reach consensus but agreed on two inferences: not to use the term 'environmental refugee'; and not to get bogged down in endless theoretical debates on definitions at the expense of practical actions that help migrants. UNU-EHS proposed a *terminology* linking triggers and responses by identifying **Environmental Emergency Migrants** who flee the worst of an environmental impact to save their lives, **Environmentally Forced Migrants** who 'have to leave' to avoid inevitable and grave consequences of environmental

T. Afifi, J. Jäger (eds.), *Environment, Forced Migration and Social Vulnerability*,
DOI 10.1007/978-3-642-12416-7, © Springer-Verlag Berlin Heidelberg 2010

degradation, and **Environmentally Motivated Migrants** who 'may leave' a steadily deteriorating environment to pre-empt the worst.

2. Is it possible to distinguish attribution of migration to interacting factors, such as environmental change and economic and social developments?

Except for the case of fleeing the consequences of swift extreme natural hazard events, the interlinkages between economic, social, and environmental factors virtually prevent the disaggregation and attribution of the various reasons which tempt people to move. Ecosystem deteriorations manifest themselves also through declining livelihoods. Thus, environmental migrants frequently claim to leave because of economic reasons but environmental change is an underlying cause.

3. Once the decision to migrate is made, does the stressor – in this case environmental change – matter? In what ways does environmentally induced migration differ from other types of migration?

Once the decision to move has been taken, the 'pull factors' attracting migrants to their destinations seem to prevail. However, the severity of the environmental stressors (push factors) may determine how precipitously one leaves. Since it is impossible, except in a few cases, to disentangle the social, economic, and environmental factors that lead to migration, it is difficult to make clear distinctions between environmentally induced and other types of migration.

Measurement and Driving Factors

4. What are the environmental 'tipping points' that can trigger migration, and when do they occur? How can we measure and predict these tipping points?

There is no environmental tipping point per se, although there is evidence of socio-environmental thresholds. Lack of coping capacities and resilience could trigger displacement. Hence, to define a tipping point, environmental events and the vulnerability of those exposed to them should be looked at simultaneously. A rapid sequence of environmental stresses seems to matter more than facing one single event, no matter how large it is.

5. How can we measure the dependence of livelihoods on environment and ecosystem services, and thereby obtain an idea of people at risk from environmentally related migration?

Since migration seems to emerge most frequently from the unfavourable interplay of environmental deterioration and the decline of ecosystem-dependent livelihood systems rather than directly from climate change or other environmental stressors, the most suitable and universal methodologies are expert interviews and statistical evaluation of questionnaires capturing the answers of potential migrants and those 'on the move'. Vulnerability assessment like the

Sustainable Livelihood Approach can be useful to identify hotspots for significant outmigration.

6. What are the interactions between poverty, environmental change, and migration? How do poverty and (social) vulnerability affect whether or not migration is a means to cope with environmental change?

A universal answer to this question is unlikely. In many cases, it is poverty alone that triggers migration. In other cases, poverty is the 'medium' through which environment affects the migration decision. People are attached to their lands, but they may leave due to livelihood vulnerabilities that are in turn influenced by the environment. If these environmental (livelihood) conditions would improve, they would rather stay. Poverty strongly influences coping capacity and resilience to deal with environmental stressors and cripples the ability to diversify livelihoods. Hence, it may act as the catalyst triggering the wish to leave. Evidence shows, however, that only those who can afford it would move. Poorest people might even be deprived of the option to migrate.

Policy Alternatives

7. Is government-initiated/managed/enforced resettlement due to environmental factors classified as environmentally induced migration? Does environmentally induced migration always have to have at least the notion of personal or group decision?

Resettlement is an enforced displacement rather than a choice and thus differs from environmentally induced migration. Experience suggests that if resettlement is to be carried out, it is essential to allow people to develop their own strategies on how they will be resettled. Environmentally induced migration always depends on a decision and most of the evidence indicates that the decisions are made at the household or perhaps community level rather than by individuals alone.

8. What legal instruments could be developed to protect/assist the different categories of environmental migrants?

The majority of conference participants favoured human rights-based approaches. The Environmental Emergency Migrants crossing a national border while seeking refuge seem to lack any type of legal recognition and basis for their plea. The participants of the conference agreed that helping alleviate the compelling reasons to migrate would effectively mitigate the need for legal recognition.

9. Presumably all environmental migration starts as internal displacement and the largest portion of this group may remain within the source country. How could recognition and assistance be given to these 'hidden' environmentally induced migrants?

Assistance frameworks conceived for internally displaced people may apply, but most effectively this category of people could be best helped through reducing their vulnerability to environmental problems by establishing fair markets, enhancing job diversification, and increasing adaptive capacity (e.g. better education, health care, land tenure). Once the people are displaced, assistance and targeted aid are needed to establish with them sustainable and culturally sensitive resettlements and to quickly develop alternative livelihoods.

Index